ANIMAL SCIENCE, ISSUES AND PROFESSIONS

# SPAWNING

# BIOLOGY, SEXUAL STRATEGIES AND ECOLOGICAL EFFECTS

# ANIMAL SCIENCE, ISSUES AND PROFESSIONS

Additional books in this series can be found on Nova's website under the Series tab.

Additional e-books in this series can be found on Nova's website under the e-book tab.

ANIMAL SCIENCE, ISSUES AND PROFESSIONS

# SPAWNING

# BIOLOGY, SEXUAL STRATEGIES AND ECOLOGICAL EFFECTS

ERICK R. BAQUEIRO CÁRDENAS
EDITOR

*New York*

For permission to use material from this book please contact us:
Telephone 631-231-7269; Fax 631-231-8175
Web Site: http://www.novapublishers.com



Additional color graphics may be available in the e-book version of this book.

**Library of Congress Cataloging-in-Publication Data**

ISBN: 978-1-63117-655-5

*Published by Nova Science Publishers, Inc. † New York*

# CONTENTS

# PREFACE

This book describes the reproductive strategies and tactics employed by bivalve and gastropod molluscs to exploit and take advantage of different environments, and their response to different environmental fluctuations.

The first chapter gives a general overview of the reproductive strategies and tactics developed through time as a means of adapting to different environments, including: morphological variations, reserve storage strategies; gametogenic effort, duration and intensity; storage of gametes, duration and intensity of spawn and post spawn; and the need and duration of a rest or recovery period.

Chapters two and three give an up-to-date overview of histological oogenesis and spermatogenesis, pointing out the variation in morphology and gametogenic processes. Chapters four and five present the different tactics adopted by different species and populations of the same species to various environments, demonstrating the plasticity of mollusc reproduction capacities, which permits them to use different environments and to thrive in changing environments.

Chapter 1 - The diversity of environments and life habits of bivalve and gastropod mollusks has promoted an equal variability of reproductive tactics within the whole range of R to K strategies, adapting to environmental conditions with diverse tactics and even strategies among populations of the same species. In this introductory chapter, author's present the general characteristics of the reproductive system and the variability of reproductive tactics within a common but diverse reproductive system.

Chapter 2 – General morphological aspects of molluscan female reproductive tract and oogenesis, in particular vitellogenesis, are reviewed in this chapter. Most molluscan are dioecious with the gonad located in the posterior part of the animal. In general, oogenesis is localized in a well-defined ovary, in where the growing oocyte is often in association with accessories cells that play an important role in its growth. Even though, in a number of species the transfer of nutrients from storage or digestive sites to the gonad has been proved. The disposition and number of accessory cells associated with single oocytes differ among molluscs and several functions have been attributed: synthesis and transfer of yolk precursor; synthesis and transfer of cytoplasmic organelles; the formation of egg envelopes; phagocytosis; hormone production; transportation of the oocyte. Oogenesis comprises a proliferative phase (premeiotic stage) followed by a growth phase (previtellogenic and vitellogenic stage). The vitellogenic stage is highly variable since the composition and organization of yolk differs among species and since different types of specialized cells may participate in the synthesis and transport of yolk or yolk precursors. The processes of yolk accumulation could be: autosynthetic, when yolk is synthesized by the oocyte itself; heterosynthetic when yolk is synthesized outside the oocyte ant then transported to it; or both. Mature oocytes display a wide range of egg envelope morphologies: primary envelopes are formed within the ovary, secondary envelopes are produced by accessory cells and tertiary envelopes formed by the accessory sex glands.

Chapter 3 – This study describes the revision of spermatogenesis and paraspermatogenesis process in gastropods and comparative sperm morphology in the most of the classes of Molluscs.

The spermatogenesis pattern including, the nuclear condensation involved granular, fibrillar, lamellar, and final homogenous electron-dense phases. Acrosome development starts with the posteriorly-located proacrosomal vesicle arising from the Golgi complex in euspermatids. This proacrosomal vesicle develops into a pre-attachment acrosome, which, together with the Golgi body, later moves towards the apex of the nucleus.

Ultrastructural studies show that spermatozoa are useful indicators of systematic position and phylogenetical relationship within the Molluscs.

The morphology of the mature sperm of Bivalves, Solenogasters, Monoplacophora, Cephalopoda and Gastropoda, given in this work, can be used to understand some aspect of the reproduction, how the shape of the sperm can be related to internal or external fertilization.

Also described is the dimorphic types of sperms in gastropods, called parasperm, and this related to the mode and the capacity of fertilization and the morphology of the parasperm also can be used as indicators of systematic position.

Paraspermatogenesis described in gastropods and the pattern of chromatin distribution in dense patches, together with the cytoplasmatic characteristics of paraspermatogonia, allow for the recognition of the apyrene line from the eupermatogonia. Later-occurring features, including the peripheral condensation of nuclear chromatin followed by nuclear invagination, the posterior breakdown into nuclear vesicles ("caryomerites"), centriole multiplication, and the synthesis of secretory products, are the most conspicuous changes in the paraspermatogenesis process.

Chapter 4 – Within the context of the reproductive biology of species, strategies would include the total reproductive model shown by a species (strategies K or R), whereas the reproductive tactics would be variations in the physiological responses that individuals show in response to changes in local environmental conditions superimposed on the typical or basic model. This paper provides a general overview of the studies conducted to date on the reproductive tactics of 14 species of marine bivalves with economic importance distributed in the northeastern region of Venezuela and the importance of environmental variables. These species belong to the Families Arcidae (*Anadara notabilis* and *Arca zebra*), Pectinidae (*Euvola ziczac* and *Nodipecten nodosus*), Ostreidae (*Crassostrea rhizophorae*), Limidae (*Lima scabra*), Mytilidae (*Perna perna*, *P. viridis* and *Modiolus squamosus*), Pteriidae (*Pinctada imbricata*), Pinnidae (*Atrina seminuda*), Donacidae (*D. denticulatus*), Psammobiidae (*Asaphis deflorata*) and Veneridae (*Tivela mactroides*). All species showed reproductive asynchrony in at least one reproductive annual half cycle, except *E. ziczac* which showed reproductive synchrony in both. These species also showed a combination of opportunistic and conservative tactics in the first and second half of annual cycle. Only *E. ziczac* contrast with this pattern, showing only opportunist tactics that coinciding with the high availability of food between January and April (first pulse of coastal upwelling). In the second reproductive half cycle in June and July, *E. ziczac* undergoes renewed active gametogenesis coincident with a second pulse of coastal upwelling. The information included in this study could be used to formulate better policies to manage fisheries resources, including the periods of fisheries closures and production strategies.

Chapter 5 – The reproductive cycles are analyzed with seven Bivalves and six gastropods to determine the variability and similarities of their reproductive cycle under diverse environmental conditions. Some species were studied at the same localities at different times and at different localities, also different species were studied simultaneously at the same locality to understand adaptive tactics to attenuate interspecific competition. Species inhabiting the same locality differ in their spawning cycle and periods of maximum activity. This study showed two gametogenic strategies as a response to the environment: populations of fast gametogenesis during a short lapse of time, and populations with a continuous gametogenesis through most of the year. The variations on the reproductive cycle of a species among different localities can be associated with environmental levels of instability. Predation and competition induce massive spawning and reduction of the spawning periods. Species inhabiting the same locality and similar habitats differ in their spawning cycle based on periods of maximum activity. Variations of one species among localities can be associated with environmental instability or variations in critical parameters with a tendency to optimize reproduction through one of several alternatives: fast gametogenesis with accumulation of mature gametes, slow constant gametogenesis with limited or no accumulation of mature gametes, constant asynchronic spawning, and synchronic spawning from limited to extended periods of time.

In: Spawning
Editor: Erick R. Baqueiro Cárdenas
ISBN: 978-1-63117-655-5


*Chapter 1*

# OVERVIEW OF BIVALVE AND GASTROPOD REPRODUCTION

*Erick R. Baqueiro Cárdenas**
Instituto Tecnológico Superior de Champotón

## ABSTRACT

The diversity of environments and life habits of bivalve and gastropod mollusks has promoted an equal variability of reproductive tactics within the whole range of R to K strategies, adapting to environmental conditions with diverse tactics and even strategies among populations of the same species. In this introductory chapter, we present the general characteristics of the reproductive system and the variability of reproductive tactics within a common but diverse reproductive system.

**Keywords:** Godan development, environmental factors, Mollusks, reproductive system

## INTRODUCTION

Reproductive cycles are a response to life strategies and environmental variations; R strategists depend on a large number of offspring with little or no parental care and little energy is spent nourishing offspring while K strategists rely on the survival of a few offspring so there is a lot of energy spent on nourishment and parental care. These strategies are generally genetically fixed for every species although environment will determine which tactics populations adopt in order to take the best advantage of prevailing conditions (Wootton, 1984).

Species have been classified into two groups, depending on the duration of their reproductive cycle: taquitictic, with short reproductive periods, and braditictic, with extended reproductive periods (Berry, 1977; Breeman, 1977; Webber, 1977). Species with a wide geographical distribution show ample variation of reproductive tactics in both duration and intensity (Bricelj & Malouf, 1980; Kennedy & Kratz, 1982; Knaub & Eversole, 1988). This has been attributed to latitude; more precisely to temperature and food availability (Bayne, 1978; Sastry, 1979; Webber, 1977; Fretter, 1984; Mackie, 1984), which has been experimentally proven by conditioning laboratory organisms of one species from different populations (Loosanoff and Davis, 1963; Bayne, 1978; Ino, 1970; Lubet & Choquet, 1971; Hines, 1979; Castagna & Kraeuter, 1981). Therefore the duration of the reproductive period is not a species attribute but a population attribute, associated with physical and biological environmental conditions.

The factors that determine population and reproductive behavior of mollusks are varied. Lucas (1965), in his thesis on the sexuality in bivalve mollusks, states that sexuality and hermaphroditism are genetically determined but strongly influenced by environmental temperature, light intensity and duration. Bricelj and Malouf (1980) found that *Mercenaria mercenaria* present different reproductive tactics: gametogenesis and spawning begin at different temperature ranges along the east coast of the United States. These patterns are preserved when individuals are transplanted to other latitudes or when they are subjected to experimental conditions, with intermediate responses from cross-breeding between organisms of different populations.

The effect of environment on reproduction was first identified by Loosanoff (1942), Shawn (1962, 1965), Kennedy and Batte (1964), Durban (1960), Ropes (1965, 1979), Langton (1987), and Walker and Heffernam (1994) among others. Cain (1974) found that for *Rangia cuneata*, salinity changes induce evacuation, and there is a negative correlation between salinity and the proportion of females.

Temperature increments are decisive in gonadal activity. However, other factors may be involved as well. Miller et al. (1981) determined the association between increased cold deep currents and the evacuation of *Argopecten gibbus*. Jones (1981) found that populations of *Spisula solidissima* and *Artica islandica* from coastal waters have a limited spawning period and restricted gametogenic activity, while populations from deep waters, though the gametogenic activity is also limited, occurs later in the year and spawning the year through. For *Panope abrupta* in British Columbia waters, Sloan and Robinson (1984) determined constant gonadal activity, with evacuation induced by increasing temperatures. Cox and Mann (1992) attributed the differences in spawning periods and fertility variations to size, magnitude of previous evacuation and parasitism.

The high morphological plasticity which allows bivalves and gastropods to occupy diverse habitats has also induced them to adopt different reproductive tactics, involving various forms of sexuality, types of eggs, embryo development, larval development, dispersal mechanisms and colonization.

## SEXUALITY AND GONADAL CYCLE IN BIVALVES AND PROSOBRANCH GASTROPODS

### *Gonadal Development in Bivalve*

In this chapter a general view of bivalve sexuality is presented. (The subject is treated in deep detail further in this book.) Most bivalves are dioecious and gonochoric. Sastry (1979) proposes four variants of pelecypod hermaphroditism: (1) Functional hermaphroditism, in which both sexes are active simultaneously, e.g., *Argopecten irradians* (Sastry, 1963), *A. circularis* (Baqueiro et al. 1981); (2) Consecutive sexuality, common among gonochoric species. Protandry is the standard development strategy, meaning that during the first period of gonadal maturation, they present male gametes, and later during the same reproductive period or on the next season, they switch to working as female, with some individuals within the population remaining as males throughout their whole life, e.g., *Mercenaria mercenaria*, *M. campechensis* (Davis and Chanley, 1955); (3) Consecutive rhythmic sexuality, species in this group reverse sex alternately from one period to another, e.g., *Ostrea* spp. (Andrews, 1979); (4) Alternating sexuality, this comprises organisms whose sex changes erratically, probably stimulated by genetic and environmental factors, e.g., *Crassostrea virginica, Chlamis distorta, Glycymeris glycymeris* (Coe, 1945; Fretter and Graham, 1964; Lucas, 1965).

Gonads originate from the mesoderm. During the rest stage, it presents as small cellular knots among the reticular connective tissue in the posterior portion of the body, in the central region of the pericardium. During maturation, the follicles grow and invade the visceral cavity. In some species the germinal tissue of the follicle differentiate into two types of cells, germ cells and nutritive cells, e.g., Mytilus; in other species the connective tissue has the nourishing function.

During spermatogenesis several generations of spermatogonia originate by mitotic division of the germ cells. Spermatogonia gives rise to first-order spermatocytes, which are free within the lumen of the follicle, forming a concentric band or isolated packets, with different development within the same follicle. The first-order spermatocytes give rise to second-order spermatocytes that quickly develop into spermatids, which will form the spermatozoa. Sperm gathers at the center of the lumen of the follicle and forms compact bundles.

Oogenesis also starts from germ cells that invade the connective tissuc then differs by mitotic division in primary oogonias. Some remain attached to the wall of the follicle while others continue their division by meiosis to give rise to second-order oogonias. Some first and second-order oogonias remain latent. Similar to spermiogenesis, some differentiate as nourishing cells, characterized by abundant granules of starch, which they gather in close proximity to the developing oogonias. The oogonias carry out a series of mitotic multiplication which later give origin to the oocytes. Oocytes begin their growth phase, which can be separated into previtellogesys and vitellogenesys. Maturing oocytes remain attached to the wall of the follicle, which gives them a characteristic pear-shape; as they mature, they move toward the center of the follicle.

## Gonadal Development in Gastropods

Among mollusks, prosobranch gastropods have the greatest diversity in reproductive biology, a product of the diversity of habitats they have colonized and different life habits. Two main groups can be distinguished by their gonadal organization: a) ditocardia, with a single gonad, which discharges through the right kidney with external fertilization, e.g., all the arqueogastropoda except for the neritids; and (b) monotocardia, also with a single gonad, which discharges by a glandular genital duct, with internal fertilization, e.g., Mesogastropoda and Neogastropoda.

The gonad is of mesoblastic origin. It develops as a proliferation of the pericardium. Primary germ cells differentiate during embryonic development (Raven, 1958). Sex chromosomes have been identified in some gastropods, such as *Turritella communis*, *Fasciolaria tulipa*, and *Nassarius mutabilis*, among others (Webber, 1977), in which the X chromosome defines masculine and Y feminine. However, it seems that several genes define sex so all gastropods can potentially change sex, which explains the presence of occasional gonochoric hermaphrodite groups (Fretter, 1984). The role of endocrine hormones in sexuality, inducing or inhibiting the expression of one sex with tentacle ganglion ablation or adding central ganglion extract, which has been called hormone CNS (Central Nervous System), has also been demonstrated.

In prosobranchs the presence of different types of sperm is common normal eupirin sperm that carries out fertilization; abnormal apirin, or without chromatic materials; and oligopirin with dispersed chromatic material. They have different functions as elements of nutrition for eupirin sperms or as assistants in locomotion of the spermatophore. An outstanding example is the formation of espermatozegumata, spermatophore capable of self-motion, that carry eupirin sperm within the female genital tract in male penis-lacking Cerithiopsis tubercularis (Webber, 1977).

## Spermatogenesis

Spermatogenesis is similar in the three groups of prosobranch gastropods, except for the spermiogenesis process (formation of spermatids and sperm) in meso and neogastrópodos where they differentiate as oligopirin and apirin sperm.

Germinal cells of the follicle epithelium undergo mitotic division, which forms the 5 μm in diameter spermatogonia cells; they remain attached to the follicle wall.

In the case of Conus, they form a syncytium with nutritive cells. Spermatogonia gives rise to spermatocytes by mitotic division; they are characterized by an increase of cytoplasmic volume and the disappearance of the nucleolus and the core wall. The first maturation or chromatin reduction is performed by meiosis. Detaching from the follicle wall, they form a concentric layer parallel to it. With the formation of the sperm cell nucleus, first-order spermatocytes pass to a resting stage, depending upon the quantity of sperm in the follicle, before continuing the second ripening (mitotic) division that gives rise to second-order spermatocytes.

The spermiogenesis process begins from spermatocytes, with the formation of sperm. Arqueogastrópodos do not form a sperm neck as the flagellum is implanted directly on the sperm head, while in meso and neogastrópodos, a neck is formed with three to five mitochondria. The shape and size ratio among head, neck and flagellum, are varied. Abnormal sperm can be formed before or after the second division, and they may contain or not chromatin material, which is usually vacuolated outside the nucleus.

### Oogenesis

The process is similar to that described for bivalves, only in meso and neogastro, numerous follicular cells contribute to the nutrition of the oocyte by either cytoplasmatic material transfer, e.g., *Monodonta* (Webber, 1977) or forming a syncytium, which incorporates both cytoplasmic and nuclear materials, e.g., *Lamellaria* (Renault, 1965). Oocyte maturation frequently occurs after sperm penetration, e.g., *Busycon, Crepidula* (Conklin, 1902 in Webber, 1977).

Occasionally, it is parthenogenetic and takes place when the oocyte is still adhered to the follicle wall. Subsequently, it loses the polar body and meiotic division occurs; there is not a chromosome reduction so diploid eggs are formed. Most of the time, the eggs mature upon contact with water or at sperm penetration. The eggs secrete substances called gynogamons and fertilizin that induce synchronous evacuation, ejaculation, and attract the sperm to the egg.

## Reproductive Cycles

The reproductive cycle may involve the life cycle of the organism, and can be initiated at the time of intercourse in the case of meso and neogastrópodos or upon gamete liberation of bivalves and arqueogastrópodos, and closed when organisms of the resulting generation are ready for reproduction. It can also be understood as the biological processes that occur within adult organisms between subsequent reproductive periods.

### Gonadal Cycle

The gonadal cycle corresponds to the sequence of events in the formation of gametes. Mackie (1984) identifies two basic types of life cycles in mollusks: semelparous, represented by annual organisms in which the generation of parents is replaced by the offsprings annually; and iteroparous, whose populations are represented by two or more generations. Both have one or more annual breeding periods, whether limited (Taquitíctics) or prolonged (Braditíctics).

These reproductive periods are controlled by internal factors, such as genetic and hormonal, and biotic external factors, such as population density, the presence of the opposite sex and predatory pressure. Abiotic factors as food, temperature, light, tidal cycle, lunar cycle, salinity and currents, are among the factors that have been correlated with fertility, gonadal maturation and evacuation induction.

The gonadal cycle description is based on microscopic observation of the state that keeps the gametes in the gonad. Lucas (1965) describes the various bivalve cycles based on cellular characteristics and compares three methodologies used practically in determining the state of maturity in bivalves. Sastry (1966) uses the degree of maturation of sex cells to define each stage. Heffernan and Walker (1989) first applied photoplanimetric techniques to determine the percentage of tissue occupied by the dominant reproductive phase. These authors define: I, a period of rest or undifferentiated gonad in which it is impossible to determine the sex; II, a period of activation or initiation of gametogenesis, corresponding to the multiplication phase (oogonia and spermatogonia); III, gametogenesis late period; IV, a period of maturity, in which the gonads are occupied almost entirely by sperm and eggs; V, a period corresponding to spawning or evacuation of the gametes; and VI, a period of post- spawn or resorption of gametes not evacuated, and the gonad enters regeneration or a rest process.

Lubet, in 1959 (in Lucas, 1965), and Galtsoff (1964) propose a number of substages while Baqueiro and Stuardo (1977) propose five phases based on the percentage of each of these engages in the gonad, including: (1) undifferentiation or rest, when it is impossible to define the sex and predominantly reticular connective tissue, (2) gametogenesis, which includes both the activation period and the late gametogenesis, (3) maturity, when the follicles are saturated with ripe gametes, (4) spawn, when follicles are partially evacuated of gametes, and 5) post spawn, in which follicles are empty except for a few non-evacuated gametes, usually found in a state of degeneration; the empty follicles are invaded by phagocytes and connective tissue starts the invasion of the follicles.

Since the determination of gonad histological stages requires training on observation and interpretation, many authors have attempted to correlate microscopic observations to macroscopic characteristics, using color and texture of the gonads. This has led to the proposition of physiological states, intended to identify or correlate to the prevailing reproductive stage; Mann (1978a), Lucas and Beninger (1985), Crosby and Gale (1990) and Rainer and Mann (1992) compared the various indices: physiologic, biochemical and volumetric, proposing the dry weight ratio of soft tissue between the volume of the shell cavity as the most reliable index for bivalves. Forgastropods, volumetric indices of the gonads, seem to be a good indicator of maturity since they do not store their reserves in the same gonad, and the increase in volume is an increase of mature gametes.

## Seasonality

Seasonality and duration of the gametogenic cycle have been mostly studied in bivalves. Bivalves and gastropods follow similar patterns defined by temperature and food availability. Usually cold-water species tend to have a short annual breeding period (Taquitíctic) and at low latitudes have extended breeding periods (Braditíctic) or several spawning periods through the year (Fretter, 1984). With respect to depth, the reproductive cycle of species from sub-littoral coastal waters, occurring between 0 and 400 m depth, is strongly influenced by environmental factors, while mollusks from the bathyal and abyssal zones of under 400 m have continuous activity with no apparent relation to external factors (Mackie, 1984).

For widespread species such as *Crassostrea virginica* and *Mercenaria mercenaria* on the East Coast of North America, it has been shown that there are breeds adapted for the conditions of temperature and salinity of their origin, and when transplanted to different regions they have different ranges of responses compared to local breeds (Bricej and Malouf, 1980; Kennedy and Krantz, 1982; Knaub and Eversole, 1988). Some species take advantage of this reproductive plasticity to colonize new areas, often displacing local species, and when there is no control by predators, they become the only species that dominates, e.g., zebra mussel in the Great Lakes of North America (Schloesser and Kovalak, 1991; Molloy et al., 1994).

It is well recognized that there is a close correlation between temperature, food availability and gonadal development cycle. Although detailed studies have demonstrated that the limiting factor in gonadal development is the presence of food, *Crassostrea virginica*, in the waters of Nova Scotia and New York, suspend gonadal development during winter. However, in the presence of food, even at temperatures close to freezing, the organisms continue with gamete formation (Bayne, 1978).

For every species and their different breeds, there is a critical temperature below which gametogenesis does not start (Bayne, 1978; Sastry, 1979; Webber, 1969; Fretter, 1984; Mackie, 1984) but once started, only the lack of food or starvation can stop it. It diminishes or increases in intensity with decreasing or increasing temperatures (Lubet and Choquet, 1971). Some species mature in the fall and store their gametes during winter so they can spawn in spring, e.g., *Spisula solidissima* in Georgia, USA (Kanti et al., 1993), or the same species may require a period of differentiation and store lipids and carbohydrates to postpone gametogenesis until the spring of the next year, e.g., *Spisula solidisima*, Princes Edwards Island, Canada (Shephton, 1987).

Reserves are stored in the follicular cells, adjacent connective tissue (Caddy, 1967; Loosanoff, 1965; Boyden, 1971), in the adductor muscle, e.g., *Argopecten* (Crenzhaw et al., 1991) or mantle and visceral mass, e.g., *Mytilus edulis* and *M. Californianus* (Hines, 1979).

## Fecundity

Bivalves and arqueogastrópodos that release their gametes to the environment have external fertilization and produce a large number of eggs, such as *Crassostrea virginica* that produces over 100 million eggs and *C. gigas* with 55 million spawned in various emissions during one season (Loosanoff, 1965). On the other hand, in those species that hatch eggs, there is an inverse relationship between the incubation time, the degree of development of the larvae when they hatch, and the number of eggs. For example, *O. lurida* released 250,000 larvae, and *O. edulis* from 100,000 to 2 million, depending on age (Hopkins, 1937 and Cole, 1941 in Andrews, 1979), or size, e.g., *Anodonta* and *Union* with 30,000 to 3,000,000 (Pennak, 1953). The lowest number of eggs is produced by those species that lay their eggs in capsules or mucus, *Nucula delphinodonta*, with 20-70 eggs per capsule (Fretter and Graham, 1964).

In those species with numerous eggs in each ovigerous capsule, most of the eggs will be consumed by the first completely developed larvae, e.g., *Buccinum undatum* with 50 to 2,000 eggs per capsule, from which only 10-30 juveniles emerge, *Natica catea* with 62 eggs, from which only 1 or 2 emerge; *Thais rustica* with more than 1,000 eggs from which emerge 400 juveniles (Webber, 1977).

The factors that regulate or stimulate evacuation may be the same or different from those that stimulate gonadal activity. Galtsoff (1954) identifies temperature as the main external agent for *Crassostrea virginica* in the U.S. East Coast; the same factor acts in most subtropical, temperate and cold-water species, such as *Mytilus edulis*, *M. californianus* (Hines, 1979) and *Venerupis japonica* (Holland and Chew, 1973). Stephner (1981) identifies salinity as the trigger for Indian species. For *Rangia cuneata*, 5 ppm salinity changes are required to induce spawning (Cain, 1974). In *Patula siliqua* the evacuation is induced by the presence of the algae *Pseudoisocrisis paradasea* (Breese and Robinson, 1981). Evacuation has been artificially induced in various bivalves using thermal stimuli, changes in salinity, chemical stimuli, electric shock, and mechanical stimulus (Ino, 1970), suggesting that in the wild there may be several factors that induce spawning.

Among gastropods, temperature seems to be the determining factor to induce evacuation. However, there may be other factors such as the type of sediment. *Strombus gigas*, *S. pugilis* and *S. costatus* require fine sand in which to lay their egg masses, and move to appropriate areas when the temperature rises (Hesse, 1972; Laughlin and Weil, 1984). *Neritina* deposit more eggs at higher salinities (Adegoke et al. in Webber, 1977) and *Littorina picta* and *L. endogenous* spawn is associated with the tidal cycle (Struhsaker, 1966).

## Fertilization

Purchon (1977) reports *Bankia gouldi* exhibit a sexual behavior of pairing and exhalant siphon intrusion of an individual in the inhalant of another with the emission of fluid sperm. For *Mercenaria mercenaria* (Castagna and Kraeuter, 1981) and *Crassostrea virginica* (Galtsoff, 1964) a cleaning action of the mantle cavity has been reported as the only reproductive behavior. *Ostrea, Union* and *Anodonta* species that incubate their eggs in their gills and mantle cavity, perform the same cleaning behavior and increase the water flow rate in the presence of sperm. There is a strong sexual dimorphism between males and females in some species of the *Montacunidae* and *Leptonidae* families, in which the much smaller male inhabits the mantle cavity or a fold of the female foot. Regarding sexual dimorphism, only *Dysnomia capsaeformis* and *D. brevidens* of the *Lamsidae* family show variations in the shell that allow the sex identification of females (Makie, 1984).

In prosobranch gastropods, different manifestations of sexual dimorphism may be present either in the shape and size of the shell (smaller for males than females), e.g., *Strombus gigas* (Alcolado, 1976), and *Ollivela biplicata* (Edwards, 1968), or they may present different correlations in the proportions of the shell according to sex, as in *Ficus subintermedia* (Arakawa and Hayashi, 1972) and *Busycon carica* (Castagna and Kraeuter, 1994). Marked size difference between sexes are shown in gastropod pests, e.g., *Entocolax* (Fretter, 1984), where the male has been reduced to a cyst of the parasite female.

A gastropod male's penis generally exhibits different accessory structures that are used for stimulating the female to spermatophore transfer. Spermatophore may be a simple aggregation of sperm in a gelatinous matrix, or form complex structures capable of movement, as in *Janthina Janthina* (Wilson and Wilson, 1956) or *Cerithiopsis tuberculata* (Fretter and Graham, 1964), a species lacking a penis.

Gastropods generally associate in groups for reproduction, as observed for *Crepidula*. Sedentary species pile on one another, where females are always larger specimens at the base. There is evidence of active recognition between the sexes, e.g., *Mitra idae* females producing mucus that attracts males (Cate, 1968). However, Reed (1995) reports a different species, *Strombus*, attempting intercourse between males and the mating between several males and one female.

## VARIATIONS IN EMBRYONIC DEVELOPMENT AND EGG TYPES

In mollusks, excepting cephalopods, segmentation of the egg is spiral dexiotropic that is clockwise, which corresponds with the twist of the visceral mass; this is also reflected in embryonic development. Raven (1958) gives a comprehensive and detailed description of embryonic development and organogenesis of mollusks. Differences have been observed between different groups of bivalves and gastropods and these differences seem to favor the success of their populations.

As mentioned, the planktonic egg with planktotrophic larval development is common among bivalves and gastropods. However, there are variations from a partial development of the larval stages in the egg or egg cases, which reduces the duration of larval stages on plankton, and therefore its dependence on environmental conditions to ova-viviparous organisms, in which embryonic development takes place inside the mother, protecting the embryo from all environmental contingencies.

Among the meiobenthos and macrobenthic bivalve species under 8 mm, it is common that the eggs are deposited in gelatinous capsules with direct development, such as *Turtolia minuita* from British waters (Oldfield, 1955) and *Abra tenuis* in Mediterranean waters (Bachelet et al., 1989), which ensures recruitment. Likewise, species from tropical and temperate waters with plaktotrophic eggs present larvae with direct development; e.g., Astarte, Macoma, Modiolaria and Modiolus (Makie, 1984). Moreover, it has been demonstrated that the viability and strength of the larvae depend on the conditions their parents were subjected to (Bayne, 1978). Larval development is abbreviated in species that hatch their eggs; e.g., *Ostrea*, *Union*; the duration of the larval period depends on the conditions the parents were subjected to.

In gastropods, the egg is protected by three walls: a) the inner wall in contact with the embryo is the wall of the egg itself, b) the second gelatinous wall; and c) the third is usually chalky or leathery in texture and is produced by the oviduct wall.

The oviduct also produces a mucoid substance that surrounds the eggs of those species that lay their eggs in masses, e.g., *Strombus* and *Natica*. These eggs can adhere to sediment particles as protection or harden into a rigid capsule; e.g., *Pomacea*.

The ovigerous capsule of *Meso* and *Neogastropoda* are made up of three layers. The two outer layers and egg protective mucus are produced by the oviduct. The degree of development that the larvae reach inside the capsule and the mechanisms of nutrition vary. Veliger larva can emerge to complete their livespans as planktotrophic; e.g., *Strombus gigas, S. pugilis, S. costatus, Littorina vincta*.

A veliger larva that does not feed may hatch to perform metamorphosis; e.g., *Acmaea testudinalis, Cerithium luctosum, Conus islandicus*; alternatively, they may present a kind of direct development, as in the genera *Busycon, Xancus, Melongena* and *Conus*. In some species, depending on the availability of food and environment to which the parents were subjected to, the embryo development may be direct or present a planktotrophic veliger larva; e.g., *Columbella rissoides* and *Brachystomia rustica* (Webber, 1977). Furthermore, as in *Planaxid suculata*, within the same capsule, embryos that develop first and have access to nutritious eggs develop directly, while the larvae that take longer to hatch, when there is no more food available, hatch as planktotrophic larva (Fretter, 1984).

The food inside the egg and the ovigerous capsule on which the embryo and larvae feed may come from different sources. Usually, species with direct development and a yolk sac have larger eggs (100-250 μm) than planktotrophic larvae. Once the larva has consumed the egg reserves, it has access to different food sources such as the protective mucous material which is rich in lipids or there may be nutritious eggs which are not fertile, e.g., *Pisania tainted, Fasciolaria tulipa, F. princeps* (Blandel, 1976). It is also common that fertile eggs with slower growth are eaten by their faster developing brothers, e.g., *Rissoa membranacea* (Rehfeldt, 1968).

The shape and arrangement of egg masses and egg capsules is specific, varying in size, number of capsules and number of eggs per ovigerous mass or capsule, according to the environmental conditions that females have been subjected to (Bandel, 1976: Hurst, 1967). Larvae that hatch from gelatinous masses do so partly by their own mechanical action and partly by autolysis of cementing material, e.g., *Strombus spp.* For egg capsules lacking a cap, juveniles scrape the wall of the capsule with their radula, e.g., *Thais hippocastanea* and *Littorina littoralis*. When there is a cap on the capsule, it is disintegrated by the action of enzymes secreted by the juveniles when they reach the end of their development (Bandel, 1976; Webber, 1977). *Pleuroploca gigantea* juveniles hatch through a cap, and they consume the egg capsule as their initial meal, making them fitter for survival than juveniles removed from the egg capsule as soon as they hatch.

The reproductive cycles are an expression of the life cycle, habitat and environmental conditions where organisms live (Mackie, 1984). Given the duration of the reproductive period, they have been grouped into two categories: (1) taquiticts, with a short and limited reproductive period, and (2) braditictics, with prolonged breeding periods. According to different authors (Bricej & Malouf, 1980; Kennedy & Kratz, 1982; Knaub & Eversole, 1988), species with a wide geographical distribution can present a range of reproductive tactics in terms of duration and intensity of the gonadal cycle (Kratz, 1982; Knaub and Eversole, 1988). These have been associated to latitude, but more closely to the temperature and food availability (Bayne, 1978; Sastry, 1979; Webber, 1979; Fretter 1984; Mackie, 1984). It has been demonstrated, when conditioning different breeds of the same species in the laboratory, that populations from different localities retain the reproductive trends present at their original sites. (Loosanoff, 1965; Bayne, 1978; Ino, 1970; Lubet and Choquet, 1971; Hines, 1979; Castagna and Kraeuter, 1981).

# Gonadal Development Stages

Every author studying gonad development may have a different interpretation of the developing stages, in particular, during the gametogenic and spawning stages. The gonad cycle has been divided from five stages, as recommended by Lucas (1965) and Baqueiro and Stuardo (1977), up to ten (Giese and Pearse, 1977, 1979; Lubet and Mann, 1987), with the basic stages being: I, Rest; II, Gametogenesis; III, Mature; IV, Spawn; V, Post-spawn. Major subdivisions have been proposed during gametogenesis and spawn, subdivisions that make it difficult for comparison between reproductive cycle descriptions. On describing the gametogenic cycle, it should be kept in mind that the gonad does not develop simultaneously. In bivalves, it starts on the pericardial wall, extending from a limited structure, as on pectinids, up to the mantle, as on mytilids. From this the necessity to apply quantitative methods to differentiate among stages can be seen. Syed-Shafe (1980) and Peterson (1986) were among the first to use planimetric measurements to characterize each stage based on a percentage of the dominant stage.

## Phase I. Undifferentiated, Or Rest

At this stage, it is important to differentiate between an undifferentiated immature organism and an adult at rest. Therefore, the initial part of any reproductive cycle study should be the determination of first maturity, to be certain that no undifferentiated immature organisms are taken for reproducers at rest.

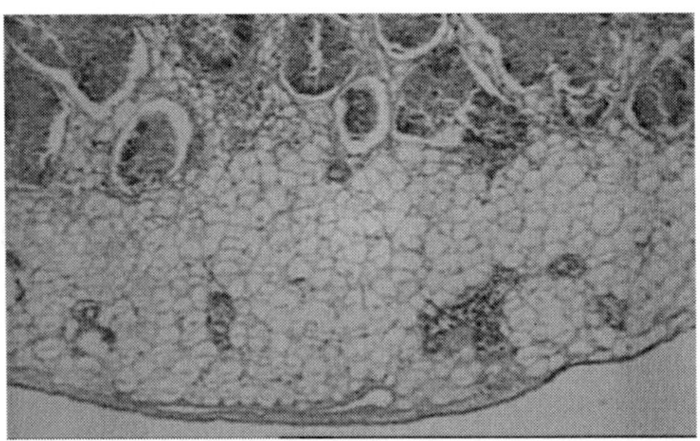

Figure 1. Undifferentiated at rest for *Busycon perversum* from Campeche Bank.

At rest, the connective tissue fills the space between the digestive gland, digestive tract, and visceral mass wall. The connective tissue presents a reticular appearance that stains pink under hematoidin-eosin with patches of blue-stained cells that form the gonadal germ cells. (Figure 1.1).

## Phase II. Gametogenesis

Among the connective tissue that fills the visceral mass from the gonad germ cells, follicles begin to form. These are characterized by a thick wall of germinal tissue, which is identified by a strong acidophilus blue stain, suggesting high chromosome activity. In females, oogonia are observed to be in contact with the germinal wall. Oocytes are the predominant stage, with reticular cytoplasm and one or two nucleolus. Mature eggs are displaced towards the center of the duct. Follicles or ducts that increase in size are observed to be anastomosed (Figure 1.2A). Piriform oocytes remain attached to the follicular wall; the cytoplasm presents a hyaline texture, and is less abundant than in the fully mature cell. The nucleus is large in relation to the amount of cytoplasm, with nuclear chromatin of a granular appearance. Two nucleoli are frequently present. Oogonia are detached from the germinal epithelium and are identified as cubic or round cells with granular cytoplasm.

In males, first and second order spermatocytes predominate, which constitute a thick layer attached to the germinal epithelium. It is possible to identify spermatids and sperm with a strong basophil stain (Figure 1.2B).

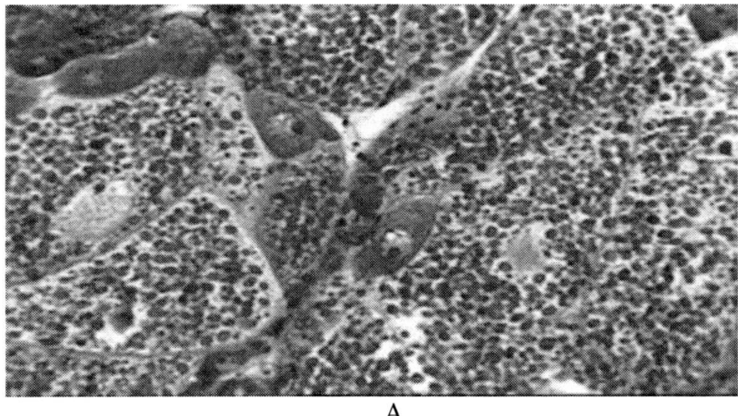

A

Figure 2. (Continued).

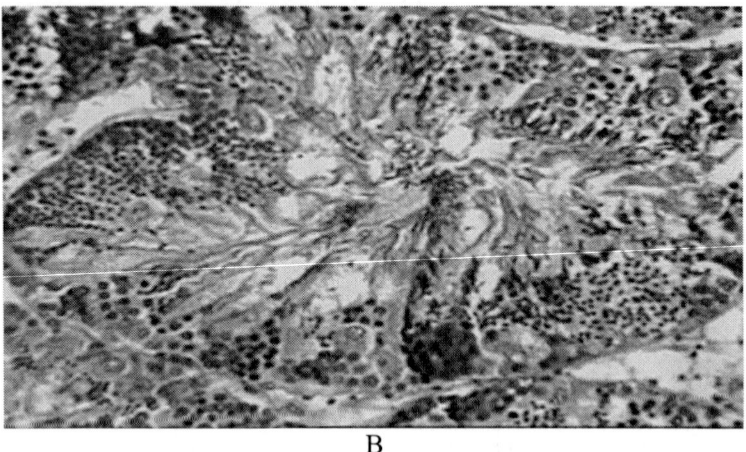

B

Figure 2. A, Female *Turbinella angulata* in gametogenesis from Seyba playa Campeche. B, Male Turbinella *angulata* in gametogenesis from Seyba playa Campeche.

## Phase III. Maturity

The detachment of the oocyte from the follicular wall indicates the overall state of maturity; these have a large nucleus with reticular chromatin and one or two nucleoli, which acquire a slightly red hue with Heidenhain's hematoxylin stain (Luna, 1968). The nuclear membrane is well defined; the cytoplasm is abundant and of a granular appearance (Figure 1.3A).

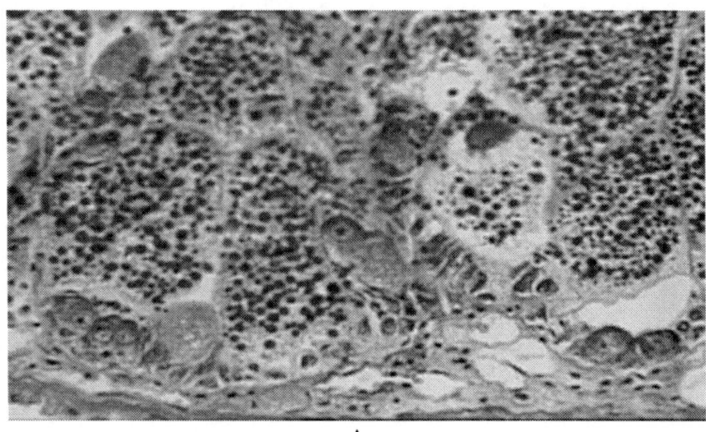

A

Figure 3. (Continued).

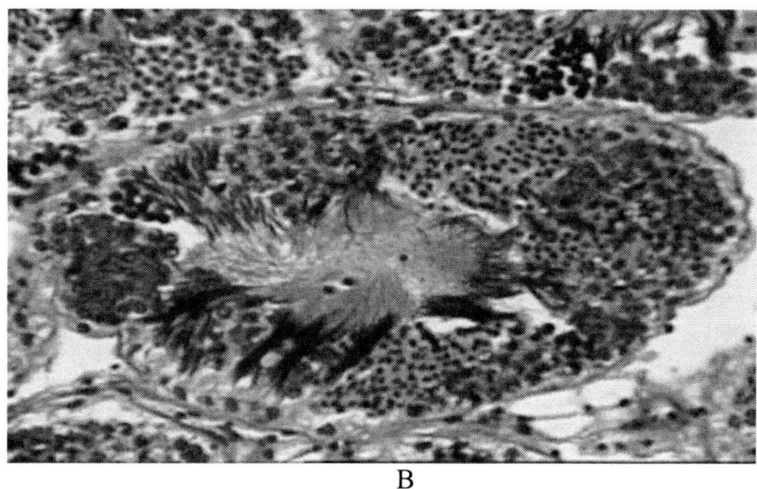

B

Figure 3. A, Mature female *Turbinella angulata* from Seyba playa Campeche. B, Mature male *Busycon perversum* from Campeche Bank.

In males, the follicles are numerous and anastomosed (Figure 1.3B), forming the follicular ducts; the interfollicular connective tissue disappears. The follicles are mostly fused, growing among the digestive gland, stomach and intestines.

At this stage, the gonad invades almost all the visceral mass and is in close contact with the digestive gland. The dominant stage in males is when mature sperm appear as bundles at the center of the follicle.

## Phase IV. Spawn

Spawn is characterized by the evacuation in varying degrees of sexual elements. The follicles are partially empty, broken and anastomosed. In females, oocytes are observed at the central portion of the ducts (Figure 1.4A). In males, follicles are partially empty, but spermiducts are saturated with sperm (Figure 1.4B). It is possible to observe some phagocytes among the connective tissue surrounding the follicles. Along the outer wall and the digestive gland, a thickening of connective tissue may be observed.

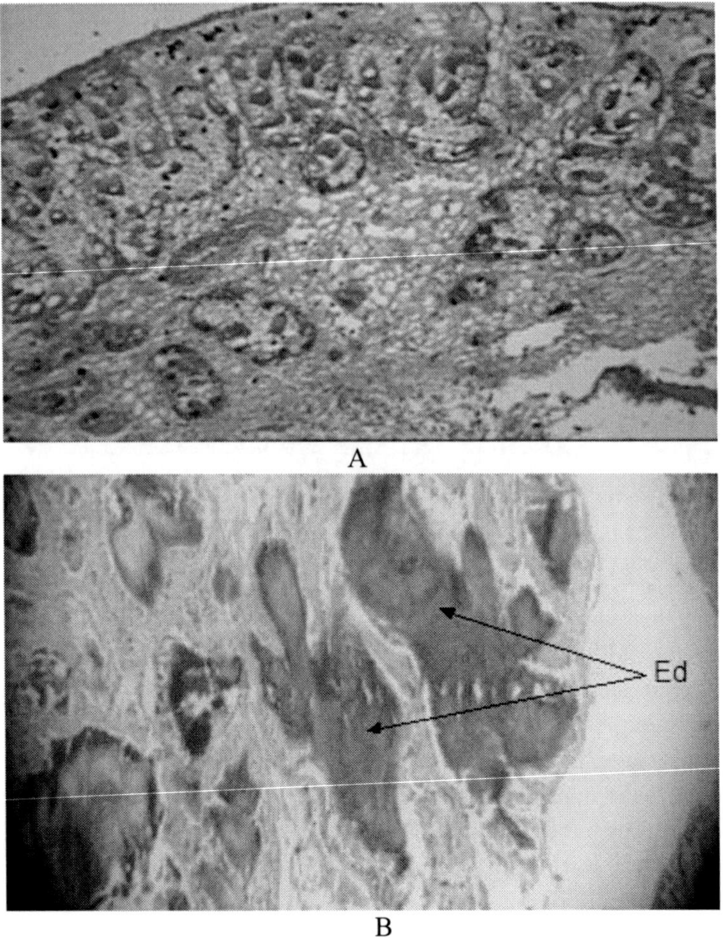


Figure 4. A, Spawning female *Turbinella angulata* from Seyba playa Campeche. B, Spawning *Fasciolaria tulipa* from Campeche bank.

## Phase V. Post Spawn

Post spawn is characterized by cytolysis and invasion of phagocytes of the gonadal tissue. This process may present different degrees of evacuation depending on whether or not gametogenesis continues during evacuation. Females present few follicles or ducts, with some broken sex cells, thereof leaving empty spaces in the lumen. Remaining eggs undergo a process of cytolysis and phagocytosis (Figure 1.5A). The follicles of males are broken; it is possible to see cells at different developing stages with signs of disintegration.

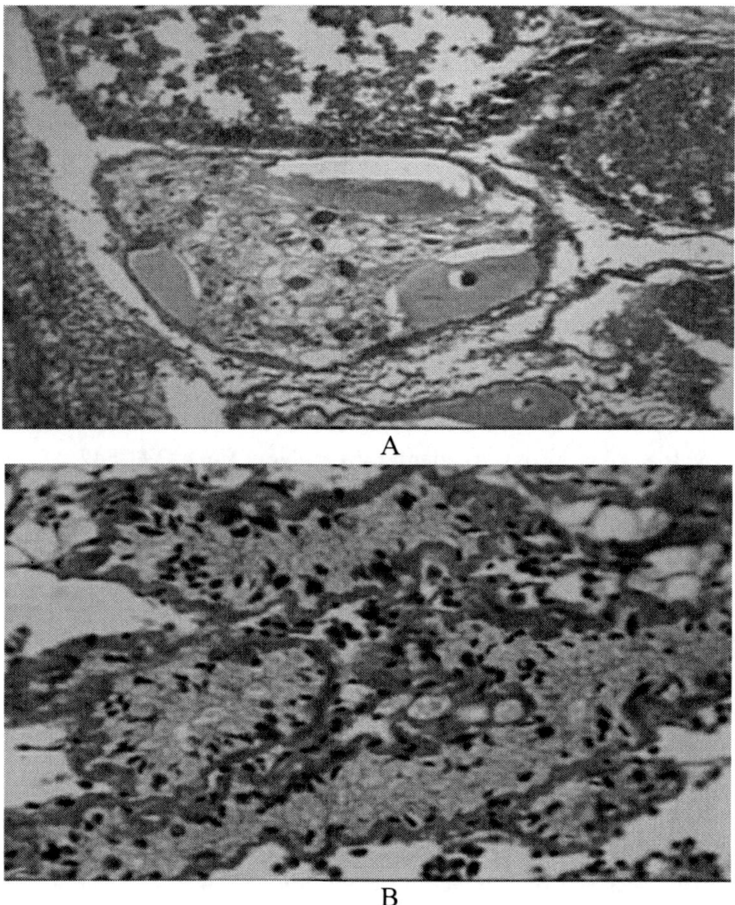


Figure 5. A, Post spawned female *Busycon perversum* from Campeche Bank. B, Post spawned male *Busycon perversum* from Campeche Bank.

The connective tissue starts to regenerate along the outer wall and around the digestive tract and digestive gland; it can be identified as a thick rod (Figure 1.5B). It also grows within the lumen of empty follicles, forming a typical reticular connective structure.

In the stages that follow post evacuation, it is possible to see traces of follicles, phagocytes, and occasionally traces of gametes that allow sex identification. Depending on environmental conditions, some populations may have a constant gametogenic activity, with partially empty follicles and active gamet maturation within the same follicle. On the other hand, if conditions are adverse, a reabsorption of the gonads may be identified by the presence of abundant germinal cells and early gametes, but few mature gametes. In females mature oocytes are usually smaller than in normal gametogenesis, and they appear broken at the center of the follicle.

## REFERENCES

Alcolado, P. M., 1976. Crecimiento y variaciones morfológicas de la concha y algunos datos biológicos del cobo *Strombus gigas* L. (Mollusca: Mesogastropoda). *Acad. Cienc. Cuba. Ser. Oceanol.* 34: 36 p.

Andrews, D. J., 1979. Pelecypoda: Ostreidae. In Reproduction of marine invertebrates, Vol. V. Molluscs: Pelecypods and lesser classes, Giese, A. C. & J. S. Pears (Eds). Academic Press. New York: 293-342.

Arakawa, A. Y. & S. Hayashi, 1972. On shell dimorphism of fig shell, Ficus subintemedia (d'Orbigny). *Venus* 31: 63-67.

Bachelet, G., J. S. Ryland, & P. A. Tyler, (Eds), 1989. Recruitment in *Abra tenuis* (Montagu) (Bivalvia, Semelidae), a species with direct development and a protracted meiobenthic phase. *Reproduction, genetics and distributions of marine organisms*: 23-30.

Bandel, K., 1976. Egg masses of 27 Caribbean Opistobranchs from Santa Marta, Columbia. *Stud. Neot. Fauna and Environ.* 11: 87-118.

Baqueiro, C. E. & J. Stuardo, 1977. Observaciones sobre la biología, ecología y explotación de *Megapitaria aurantiaca* (Sow., 1931), *M. squalida* (Sow. ) y *Dosinia ponderosa* (Gray, 1938) de Zihuatanejo e isla Ixtapa, Gro., México. An. Cent. Cienc. del Mar y Limnol. Univ. Nal. Autón. , México 4 : 161 - 208.

Baqueiro, C. E., I. Peña R. & J. A. Masso, 1981. Análisis de una población sobre-explotada de *Argopecten circularis* (Sowerby, 1835) en la ensenada de La Paz, B. C. S., México. *Ciencia pesquera. Inst. Nal. Pesca Depto.* Pesca. México, 1 : 57 - 65.

Bayne, B., 1978. Reproduction of bivalve molluscs under environmental stress. In: *Physiological ecology of estuarine organisms*. J. F. Vernberg (Ed) the Belle W. Baruch library in marine science, New York : 259-278.

Beeman, R. P., 1977. Gastropoda: Oisthobranchia. In: *Reproduction of Marine Invertebrates,* Vol. 4, Cap. 2. Giese A. C. & J. S. Pearse (Ed), Acadmic Press. New York 115 - 180.

Berry, A.J., 1977. Gastropoda: Pulmonata. In: *Reproduction of Marine Invertebrates,* Vol. 4, Cap. 3. Giese A.C. & J.S. Pearse (Ed), Acadmic Press. New York, 181 – 241.

Boyden, C. R., 1971. A comparative study of the reproductive cycles of the cockles *Cerastoderma edule* and *C. glaucum*. *J. Mar. Biol. Ass. U. K.* 51: 605-622.

Breese, W. P. & A. Robinson, 1981. Razor clams, *Siliqua patula* (Dixon): gonadal development, induced spawning and larval rearing. *Aquaculture* 22: 27-33.

Bricelj, M. V. & R. E. Malouf, 1980. Aspects of reproduction of hard clams (*Mercenaria mercenaria*) in Great South Bay, New York. *Proc. Nat. Shellf. Ass.* 70, 216 - 229.

Caddy, J. F., 1967. Maturation of gametes and spawning in *Macoma baltica* (L.). *Can. J. Zool.* 45: 955-965.

Cain, T. M., 1974. Reproduction and recruitment of the brackish water clam *Rangia cuneata*. *Fish. Bull.* 73 : 412-430.

Castagna, M. & J. N. Kaeuter, 1981. Manual for growing the hard clam *Mercenaria*. Spec. Vir. Inst. Mar. *Sci. Rep. Apl. Mar. Sci. and Ocean Enger.* 249 : 110 pp.

Castagna, M. & J. Kraeuter, 1994. Age, growth rate, sexual dimorphism and fecundity of knobbed whelk *Busycon carica* (Gmelin, 1791) in western Mid-Atlantic lagoons systems, Virginia, *J. Shellf. Res.* 13 : 581 - 586.

Cate, J. M., 1968. *Mating behavior in Mitra idae Melvill*, 1893. Veliger 10: 247-252.

Coe, W. R., 1945. Development of the reproductive cycle and variations in sexuality in Pecten and other pelecypod molluscs. *Trans. Conn. Acad. Sci.* 36, 673-700.

Cox, C. & R. Mann, 1992. Temporal and spatial changes in fecundity of Eastern oyster, *Crassostrea virginica*, (Gmelin 1791) in the James river, Virginia. *J. Shellfr.* 11 : 49 - 54.

Crenzhaw, J. W., P. B. Hefferman & R. L. Walker, 1991. Heritability of growth rate in the southern bay scallop, *Argopecten irradians* concentricus (Say, 1822). J. *Shellf. Res.* 10 : 55-63.

Crosby, P. M. & L. D. Gale, 1990. A review and evaluation of bivalve condition index methodologies with a suggested standard method. *J. Shellf. Res.* 9 : 233-238.

Davis, H. C. & P. E. Chanlely, 1955. Spawning and egg production of oysters and clams. *Proc. Nat. Shellf. Ass.* 46, 40-51.

Durban, M J., 1960. The evolution of stability in marine environments, natural selection at the level of the ecosystem. In: Readings in Marine Ecology, J. W. Nybakken (Ed), *Harper and Row,* : 469 - 476.

Edwards, D. C., 1968. Reproduction in *Olivela biplicata*. Veliger 10: 297-304.

Fretter, V. & A. Graham, 1964. Reproduction. In: *Physiology of Mollusca.* Wlburn, K. M. & C. M. Yong (Eds), Vol. 1. Academic Press, New York: 127-164.

Fretter, V., 1984. Prosobranchs. In: The Mollusca, Wilburn, M K. (Ed), Vol 7. *Reproduction,* Tompa, A. S., N. H. Verdonk & J. A. M. den Biggelaar (Eds). Academic Press, Orlando Florida, Cap. 1: 1-45.

Galtsoff, S. F., 1964. The American Oyster. Fishr. Bull. 64: 980 pp.

Giese, A. C. & J. S. Pearse, 1977. *Reproduction of Marine Invertebrates. Molluscs: Gastropods and Cephalopods*. Vol. IV. Academic Press, New York: 368 pp.

Giese, A. C. & J. S. Pearse, 1979. *Reproduction of Marine Invertebrates. Molluscs: Pelecypods and Lesser classes*. Vol. V. Academic Press, New York: 368 pp.

Hefferman, P. B. & R. L. Walker, 1989. *Gametogenic cycle of three bivalves in Wassaw sound,* Georgia. III Geukensia demissa (Dillwyn, 1817). J. Shellf. Res. 8 : 327-334.

Hesse, O. K., 1972. An ecological study of the queen conch *Strombus gigas*. Tesis M. Ci. Connecticut Univ.: 107 pp.

Hines, A. H., 1979. Effects of a thermal discharge on reproductive cycles in *Mytilus edulis* and Mytilus californianus (Mollusca, bivalvia). *Fish. Bull.* 77 : 125-132.

Holland, D. A. & K. K. Chew, 1973. Reproductive cycle of the manila clam (*Venerupis japonica*) from hood canal, Washington. *Proc. Nat. Shellf. Ass.* 64: 65-72.

Hurst, A., 1976. *The egg masses and veligers of thirty northeast Pacific opistrobranches*. Veliger 9: 255-288.

Ino, T., 1970. Controlled breeding of molluscs. Indo Pacific Fishr. Coun. F. A. O. IPFC/070/SYM 35: 24 pp.

Jones, S. D., 1981. Reproductive cycle of the Atlantic surf clam *Spisula solidissima*, and the ocean quahog *Artica islandica* off New Jersey. *J. Shellf. Res.* 1, 23-32.

Kanti, A., P. B. Heffernan & R. L. Walker, 1993. Gametogenic cycle of the southern surfclam, *Spisula solidissima* similis (Say, 1822) from St. Catherines sound, Georgia. *J. Shellf. Res*. 12 : 255-261.

Kennedy , V. S. & L. B. Krantz, 1982. Comparative gametogenic and spawning patterns of the oyster *Crassostrea virginica* (Gmelin) in central Chesapeake bay. *J. Shellf. Res.* 2 : 133-140.

Kennedy, A. V. & H. I. Batte, 1964. Cyclic changes in the gonad of the american oyster *Crassostrea virginica* (gmelin). *Can. J. Zool.* 42 : 305 - 321.

Knaub, R. S. & A. G. Eversole, 1988. Reproduction of different stocks of *Mercenaria mercenaria. J. Shellf. Res.* 7 : 371-376.

Langton, W. R., 1987. Fecundity and reproductive effort of sea scallops *Placopecten megallanicus* from the Gulf of Maine. *Mar. Ecol. Prog. Ser.* 37: 19 - 25.

Loosanoff, V. L. & H. C. Davis, 1963. Rearing of bivalve mollusks. In: *Advances in Marine Biology. Russell,* F. S. (Ed) Academic Press, New York 1: 2 - 136.

Loosanoff, V. L., 1942. Seasonal gonadal changes in the adult oysters, *Crassostrea virginica*, of Long Island Sound. *Biol. Bull.* 82, 195 - 206.

Loosanoff, V., 1965. Gonad development and discharge of spawn in oysters of Long Island Sound. *Biol. Bull. Mar. Biol. Lab. Woods Hole* 129, 546-561.

Lubet, P. & C. Choquet, 1971. *Cycles et rythmes sexuales chez les mollusques bivalves et gastropods.* Haliotis 1 : 129-149.

Lubet, P. & R. Mann, 1987. Les Diferentes modalites de la reproduction chez les Mollusques bivalves. Haliotis, 16: 181 - 195.

Lucas, A., 1965. Recherche sur la sexualité des mollusques bivalves. Tesis doctoral Fac. Sci. Univ. Rennes, Francia: 136 pp.

Lucas, A. & P. G. Beninger, 1985. The use of physiological condition indices in marine bivalve aquaculture. *Aquaculture* 44: 187-200.

Luna, G. L., 1968. *Manual of Histologic staining methods of the Armed Forces Institute of Pathology*. Tercera Ed. McGraw-Hill Book Co. New York: 258 pp.

Mackie, G. L., 1984. Bivalves. In: The Mollusca. Wilbur, K. M. (Ed). Vol. 7. *Reproduction T*ompa, S. A., N. H. Verdonk & J. A. M van den Biggelaar, Ac. Press. Cap. 5: 305-418.

Mann, R., 1978 a. Comparison of morphometric, biochemical and physiological indexes of condition in marine bivalve mollusks. In: *Energy and environmental stress in aquatic systems.* Throp, J. H. & J. L. W. Gibbons (Eds) D.O.E. Symposium ser.: 484-497.

Molloy, P. D., J. Powell & P. Ambrose, 1994. Short-term reduction of adult zebra mussels, *Dreissena polymorpha*, in the Hudson River near Catskill,

New York: An effect of juvenile blue crab Callinectes sapidus predation?. *J. Shellf. Res.* 13 : 367-372.

Oldfield, E., 1955. Observations on the anatomy and mode of life of *Lasaea rubra* (Montagu) and *Turtonia minuta* (Fabricus). *Proc. Malacol. Soc.* London. 31 : 236-249.

Pennak, W. R., 1953. Fresh-water invertebrates of the United States. *Pelecypoda* (Clams, mussels). Cap 37. The Ronald Press Co. New York, : 694 - 726.

Peterson, C. H., 1986. Quantitative allometry of gamete production by *Mercenaria mercenaria* into old algae. *Mar. Ecol. Prog. Ser.* 29: 93 - 97.

Purchon, R. D., 1977. *The Biology of the Mollusca.* Seg. ed. Pergamon Press, New York : 312 pp.

Rainer, S. J. & R. Mann, 1992. A comparison of methods for calculating condition index in eastern oysters, *Crassostrea virginica* (Gmelin, 1791). *J. Shellf. Res.* 11 : 55-58.

Raven, P. C., 1958. Morphogenesis: The analysis of Molluscan development. *Int. Ser. Monog. on Pure and Applied Biol. Div.: Zool.* Pergamon Press, London: 311 pp.

Reed, E. S., 1995. Reproductive anatomy and biology of the genus *Strombus* in the Caribbean: I Males. *J. Shellf. Res.* 14: 325-330.

Rehfeldt, E., 1968. Productive and morphological variations in the prosobranchs "*Rissoa membranacea*". *Ophelia* 5: 157-173.

Renault, L., 1965. Observations sur l'ovogenèse et sur les cellules nourriciéres chez *Lamellaria persipicua* (L.) (Mollusque Prosobranche). *Bull. Mus. Natl. Hist. Nat.,* Paris 37: 282-284.

Ropes, J. W., 1979. Shell length at sexual maturity of surf clams, *Spisula solidissima*, from an inshore habitat. *Proc. Nat. Shellf. Ass.* 69: 85 - 88.

Ropes, J., 1965. Reproductive cycle of *Mya arenaria* in New England. *Biol. Bull.* 128 : 315 - 327.

Sastry, N. A., 1963. Reproduction of the bay scallop, *Aequipecten irradians* Lamarck. Influence of temperature on maturation and spawning. *Biol. Bull.* 125: 146-153.

Sastrty, N. A., 1966. The development and external morphology of pelagic larval and post-larval stages of the bay scallop, *Aequipecten irradians concentricus* Say, reared in the laboratory. *Bull. Mar. Sci.* 15: 417-435

Sastry, N. A., 1979. Pelecypoda (Excluding Ostreidae). In: *Reproduction of Marine Invertebrates,* Giese, A. C. & J. S. Pearse, (Ed), Academic Press, Vol. 5: 113 - 292.

Shawn, W. N., 1962. Seasonal gonadal changes in female soft shell clams *Mya arenaria*, in Tred Avon river, Maryland. *Proc. Natn. Shellfish,* Assc. 52 : 121 - 132.

Shawn, W. N., 1965. Seasonal gonadal cycle of the male surf clam, *Mya arenaria*, in Maryland. U. S. Fish. Wildl. *Serv. Spec. Sci. Rep. Fish*. 508 : 5p.

Shephton, T. W., 1987. The reproductive strategy of the Atlantic sur clam, *Spisula solidissima*, in Prince Edwards island, Canada. *J. Shellf. Res*. 6, 97-102.

Shloesser, D. W. & W. Kovalka, 1991. Infestation of unionids by *Dressena polymorpha* in a power plant in Lake Erie. *J. Shellf. Res*. 10 : 355 - 360.

Sloan, N. A. & S. M. Robinson, 1984. Age and gonad development in the geoduck clam *Panope abrupta* (Conrad) from Southern British Columbia, Canada. *J. Shellf. Res.* 4, 131 - 137.

Stephner, D., 1981. Induction of spawning four species of bivalves of the Indian coastal waters. Aquaculture 25: 153-159.

Struhsaker, J. W., 1966. Breeding, spawning, spawning periodicity and early development in the Hawaiian *Littorina*: *L. pintada* (Wood), *L. picta* Philippi, and *L. scabra* (Linné). *Proc. Malacol.* Soc. London. 37: 137-366.

Syed-Shafe, M., 1980. Quantitative studies on the reproduction of black scallop, *Chlamys varia* (L.) form Lanveoc area (Bay of Brest). *J. Exp. Mar. Biol. Ecol.* 42 : 171-186.

Walker, L. R. & P. B. Heffernam, 1994. Temporal and spatial effects of tidal exposure on the gametogenic cycle of the northern quahog, *Mercenaria mercenaria* (Linnaeus, 1758) in coastal Georgia. *J. Shellf. Res.* 13 : 479 - 486.

Webber, H. H. & A. C. Giese, 1969. Reproductive cycle and gametogenesis in the black abalone *Haliotis chacherodii* (Gastropoda: Prosobranchia). *Mar. Biol.* 4 : 152-159.

Webber, H. H., 1977. Gastropoda: Prosobranchia. In: Reproduction of Marine Invertebrates, Vol. IV, Cap. 1, *Molluscs: Gastropod and Cephalopods*. Acadmic Press. New York, 1 - 98.

Wilson, W. P. & M. A. Wilson, 1956. A contribution to the biology of *Janthina janthina* (L.). *J. Mar. Biol. Assoc.* U. K. 35: 291-305.

In: Spawning
Editor: Erick R. Baqueiro Cárdenas
ISBN: 978-1-63117-655-5


*Chapter 2*

# MOLLUSCAN OOGENESIS

***Florencia Arrighetti***
Biología de Invertebrados Marinos. DBBE. Facultad de Ciencias Exactas y Naturales. Universidad de Buenos Aires, Argentina.
IBBEA, CONICET- UBA
CONICET- Museo Argentino de Ciencias Naturales "Bernardino Rivadavia"

## ABSTRACT

General morphological aspects of molluscan female reproductive tract and oogenesis, in particular vitellogenesis, are reviewed in this chapter. Most molluscan are dioecious with the gonad located in the posterior part of the animal. In general, oogenesis is localized in a well-defined ovary, in where the growing oocyte is often in association with accessories cells that play an important role in its growth. Even though, in a number of species the transfer of nutrients from storage or digestive sites to the gonad has been proved. The disposition and number of accessory cells associated with single oocytes differ among molluscs and several functions have been attributed: synthesis and transfer of yolk precursor; synthesis and transfer of cytoplasmic organelles; the formation of egg envelopes; phagocytosis; hormone production; transportation of the oocyte. Oogenesis comprises a proliferative phase (premeiotic stage) followed by a growth phase (previtellogenic and vitellogenic stage). The vitellogenic stage is highly variable since the composition and organization of yolk differs among species and since different types of specialized cells may participate in the synthesis and transport of yolk or yolk precursors. The processes of yolk accumulation could be: autosynthetic, when yolk is synthesized by the oocyte itself; heterosynthetic when yolk is synthesized outside the oocyte ant then transported to it; or both. Mature oocytes display a wide range of egg envelope morphologies: primary envelopes are formed within the ovary, secondary envelopes are produced by accessory cells and tertiary envelopes formed by the accessory sex glands.

## General Aspects of the Female Reproductive Tract

Understanding the life history pattern of an organism should lead to a better knowledge of the selective forces that shape the evolution of life histories. In this sense, many life history traits are constrained by ovarian structure and by the vitellogenic mechanisms that determine the quantity, quality and rate of yolk incorporation into the egg during oogenesis (Eckelbarger, 1994).

Primitively, molluscs are dioecious, with a pair of gonads displaced towards the anterior portion of the body (Figure 1). When mature molluscs discharge their developing gametes to the outside, either through the nephridial plumbing or through separate ducts. Fertilization is external and development is indirect although many molluscs have various means of internal fertilization.

Most aplacophorans are hermaphroditic species with a pair of gonads although some are dioecious with single or paired gonads (Figure 2) connected to the pericardial chamber by means of gonopericardial ducts. Monoplacophora are dioecious and possess two pair of gonads that release their gametes in each gonoduct connected to one of the pairs of nephridia, fertilization is external (Figure 3). The polyplacophora possess one pair of gonads which shows the tendency to fuse to one unpaired gonad situated medially in front of the pericardical cavity (Figure 4). Gametes are transported to the outside by two separate gonoducts that opens directly into the mantle cavity. In some species fertilization is external although often occurs in the mantle cavity. In most bivalves sexes are usually separate with a pair gonad closely invested with the viscera and with each other (Figure 5). Some species of oysters, scallops, cockles, and others are hermaphroditic. Most of the bivalves, with exception of the Pectinidae, lack a discrete gonad. Instead, gonad tissue arises from the mesoderm and develops gradually and diffusely around the digestive gland (Figure 5). Gametes are discharged into the mantle cavity separately form the nephridiopores. Fecundation is usually external although in some freshwater bivalves is internal and brood their embryos. Gastropods may be dioecious or hermaphroditic with only a single gonad usually coiled within the visceral mass (Figure 6). The gonoduct is always developed in association with the right nephridium, the isolation of the reproductive tract form the excretory system allowed the great variety of reproductive and developmental pattern found in gastropods. Scaphopods are dioecious with an unpaired gonad in the posterior region of the body. Gametes are discharge throughout the right nephridium (Figure 7). Sexes are separated in cephalopods, with a single gonad in the posterior region of the visceral mass (Figure 8). In male various glands assist in packaging the sperm into spermatophores, which are stored in a sac that opens into the mantle cavity.

## Extragonadal Tissues

Extragonadal tissues producing vitellogenins do not occur in molluscs. However, in certain groups, some specializations have been described that enable oocytes to obtain sufficient nutrients for the vitellogenic process (Jong-Brink et al., 1983). In the chiton Sypharochiton septentriones during vitellogenesis a special vascular system develops around the ovary. It was observed, at the onset of vitellogenesis, an increase in the surface area of the plasma membrane exposed to the blood in the sinuses and the presence of micropinocytotic vesicles in the oocytes, both observations are indicative that some protein yolk is derived from material in the blood stream (Selwood, 1968). In bivalves, the amount of this tissue varies throughout the reproductive cycle, and it probably serves as an energy reserve for the developing gametes. The transfer of nutrients from storage or digestive sites to the gonad has been studied in a number of species. Le Pennec et al. (1991), based on anatomical, ultrastructural and histochemical data, proposed for Pecten maximus a transfer of nutrients from the intestine to the developing gametes. Subsequently, the use of ferritin, an iron-containing transfer protein, confirms this proposal (Beninger et al., 2003)

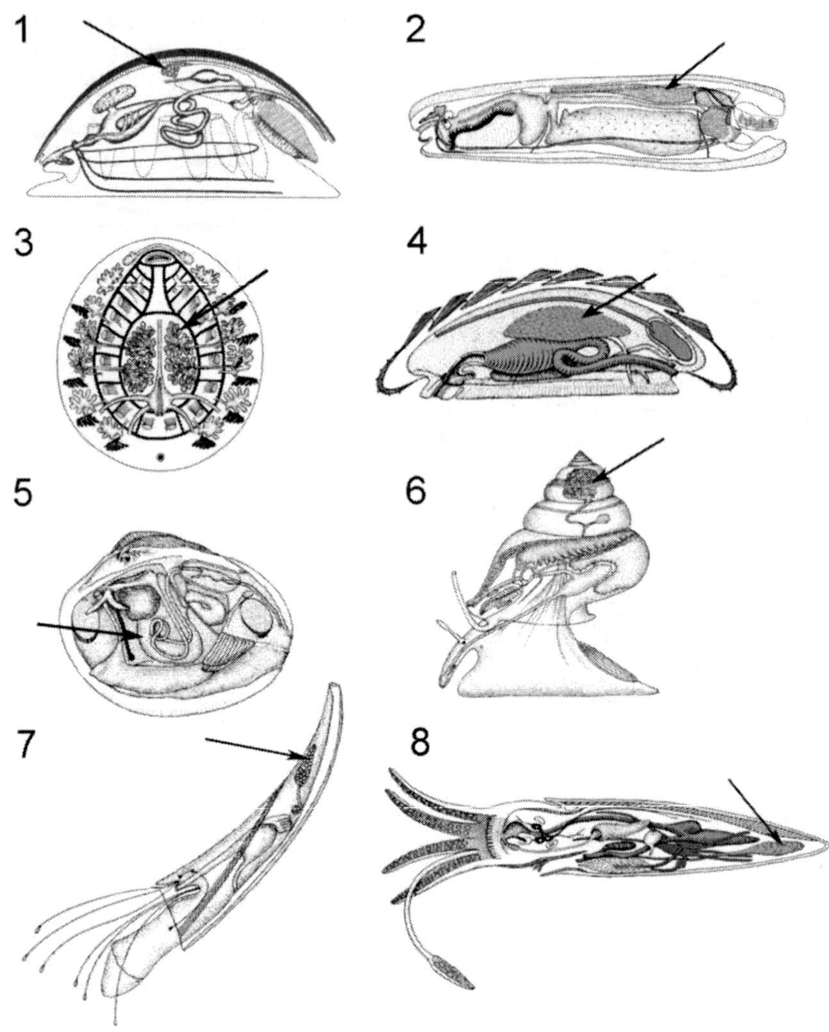


Figures 1-8. General anatomy of the different classes of Mollusca, showing the position of the gonad (arrow). 1. Generalized mollusca. 2. Aplacophora. 3. Monoplacophora. 4. Polyplacophora. 5. Bivalvia. 6. Gastropoda. 7. Scaphopoda. 8. Cephalopoda. Modified form: Ruppert and Barnes, 1996 (1); Hickman et al., 2001 (4, 7); Barnes and Barnes, 2002 (2, 3, 5, 6, 8).

In bivalves, the name inter-acinal connective tissue is widely used to describe the tissues found between the acini. In *Crassostrea virginica*, the inter-acinal spaces are largely filled with vesicular connective tissue cells (VCT) that contain lipid droplets and glycogen (Eckelbarger and Davies,

1996). These cells are also found in *Mytilus edulis* but only contain glycogen (Pipe, 1987a). Another storage cell type is present in this species, the adipogranular cells (ADG), which are composed predominantly of protein granules, lipid droplets and glycogen (Pipe, 1987a). The gametogenic condition determines the relative proportion of storage tissue to germinal cells, usually forming an annual or biannual cycle that depends upon the location of the population (Lowe et al., 1982; Pipe, 1987a). This partitioning of nutrient reserves into two storage cell types is useful for animals undertaking gametogenesis during winter months, where the availability of food nutrients is low. Pipe (1987a) demonstrate for *M. edulis,* using quantitative histochemistry, that ADG cells provides a rapid source of nutrients when food levels are inadequate while VCT cells are useful in controlling the long term seasonal demands on nutrients.

In the pulmonate snail *Lymnaea stagnalis* vitellogenic oocytes are primarily located in vitellogenic areas, opposite the diverticula of the digestive gland. This portion of the ovotestis is favorable for the uptake of nutrients from the digestive system as was demonstrated by Bottke et al. (1982). In *Octopus vulgaris*, the oocytes are individually supplied with a capillary system. Vitellogenesis occurs at the end of the life cycle, and this event coincides with an active catabolism of muscle protein, which leads to an increase of amino acid pools in the blood. These probably serve as the main source for protein yolk synthesis (Jong-Brink et al., 1983) and the direct use of protein as an energy reserve is due to the lack of major glycogen and lipid reserves in cephalopod tissues (Rosa et al., 2004).

## ACCESSORY CELLS

In this chapter, the term accessory cells will be used to refer those cells of somatic origin that accompanying the oocyte during it development. Accessory cells (also named follicle cells, auxiliary cells, companion cells, supportive cells) appear nearly universally in invertebrate gonads, and often contain abundant proteosynthetic organelles that are probably involved in yolk precursor production (Wourms, 1987; Eckelbarger, 1994). Other functions have been attributing to accessory cells: synthesis and transfer of cytoplasmic organelles; the formation of egg envelopes; phagocytosis; hormone production; transportation of the oocyte.

The disposition and number of accessory cells associated with single oocytes differ among molluscs. Jong-Brink et al. (1983) distinguish three types of arrangements. The first type of arrangement is characterized by a changing relationship between oocytes and accessory cells during the growth of the oocyte. In early stages of oocyte development, the oocyte becomes completely surrounded by a limited number of accessory cells, which plays a role in the exchange between the germinal cells and the surround (Figure 9) (Griffond and Gomot, 1979). In later stages, accessory cells detach form the apex of the oocyte which now bulges freely into the lumen of the acinus. In some species of bivalves the oocyte keeps contact with the acinar wall by a stalk (Figure 10) (Pipe, 1987b). In the second type, the oocyte becomes surrounded by a small and distinct number of accessory cells, keeping contact with the basement membrane of the acinar wall. In the third type, the developing oocyte becomes completely surrounded by an increasing number of accessory cells, which often form a syncytium, as the case of *O. vulgaris* and *S. septentriones*, but in the later the oocyte keeps contact with the acinar wall by a thin stalk (Figure 11).

In the poliplacophora *S. septentriones* each oocyte is surrounded by a layer of accessory cells (also named follicle cells), which are joined to each other by desmosome-like structures. The accessory cells of young oocytes are flattened squamous-like cells with an elongate nucleus. As the oocyte develops there is an increase in the accessory cells volume and in association with this increment is the production of the secondary coat. The accessory cells do not appear to be directly involved in oocyte nutrition as was suggested by Selwood (1968). Later, Selwood (1970) demonstrate that the main role of this accessory cells was the deposition of a spiny egg hull around each oocyte.

In most of the bivalves studied, the accessory cells are intra-acinal and usually are associated with each oocyte, at least during early oogenesis as the case of *M. edulis* (Pipe, 1987), *Pinna nobilis* (De Gaulejac et al., 1995). As an exception, in the scallop *P. maximus*, each developing oocyte is closely associated with an accessory cell from the beginning of the diplotene stage to the final stage of the vitelline coat (Dorange and Le Pennec, 1989). The accessory cells in *M. edulis* have de ability for uptake, turnover and synthesis of material (Pipe, 1987b) and the presence of desmosome-like gap junctions indicate that exchange of small molecules and ions does take place between accessory cells and oocytes. The same structures were observed by Eckelbarger and Davies (1996) in *C. virginica*, hence accessory cells could also function to control the environment around the oocytes by regulating the flow of small molecules and ions to oocytes. In *C. virginica* is more likely that the major source of precursors originates in the VCT cells that surrounds the ovarian acinus.

In the gastropod *Viviparus viviparus* the storage of glycogen and lipids detected in accessory cells and the close relationship observed between this cells and oocytes suggest an important role in the nutrition of the oocyte. Moreover, the presence of lysosomes and residual bodies in accessory cells indicates an active participation in phagocytosis of the residual oocytes (Griffond and Gomot, 1979). Some particular arrangement was observed in the volutid *Adelomelon beckii* in where oocytes are always surrounded by a great number of accessory cells. During vitellogenesis, gap junctions are observed between accessory cells and oocytes suggesting the exchange of small molecules and ions (Arrighetti and Penchaszadeh, 2010).

Several types of accessory cells are associated with tertiary egg envelope formation and the nutrition of developing eggs in gastropod (see section 5).

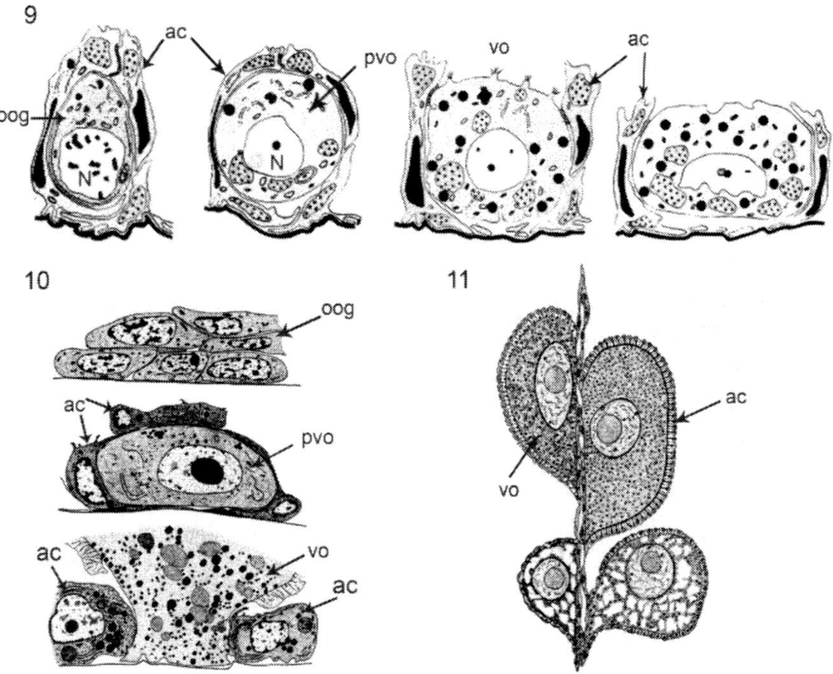


Figures 9-11. Interrelationship between oocytes and accessory cells during oogenesis: 9. In the gastropod *Viviparus viviparous*. 10. In the bivalve *Mytilus edulis*. 11. In the chiton *Sypharochiton septentriones*. *Oog* oogonia, *ac* accessory cell, *pvo* previtellogenic oocyte, *vo* vitellogenic oocyte. Modified from: Griffond and Gomot, 1979 (9); Pipe, 1987 (10); Selwood, 1968 (11).

# OOGENESIS

Oogenesis and spawning are controlled by both endogenous and exogenous factors. Environmental factors (e.g., temperature, photoperiod, oxygen content of the water) play an important role in influencing ovipository activity. The influence of these factors were studied in several species of chitons (Nagabhushanam and Deshpande, 1982; Currie, 1990); bivalves (Chung and Ryou, 2000; Chung, 2008; Son and Chung, 2009); gastropods (Giese and Pearce, 1974; Hayashi, 1980; Martel et al., 1986; Fretter and Graham, 1994; Giménez and Penchaszadeh, 2002; Ilano et al., 2003; Arrighetti and Penchaszadeh, 2010) and cephalopods (Rodríguez-Rúa et al., 2003; Ortiz et al., 2011). All of the environmental factors mentioned above causes physiological changes within the individual that are regulated through endocrine control (for a review see Schroeder, 1987).

Oogenesis starts with the appearance of a distinct population of primordial germ cells who differentiate as female gametes and proliferate to produce oogonia (Wourms, 1987). Primary oogonia (or proliferating oogonia) are small cells characterize by a high nuclear-cytoplasmic ratio. These cells undergo a series of mitotic divisions, producing a number of secondary oogonia (or terminal oogonia). This proliferative phase finish when the secondary oogonia are transformed into primary oocytes. The period of oocyte differentiation has been divided into premeiotic, previtellogenic and vitellogenic phases.

In most of the molluscs, complete meiosis of the female gametes is delayed until spawning and fertilization takes place (Beninger and Le Pennec, 2006). For example, in scallops, the mature oocyte is blocked in the first metaphase of meiosis; the remaining meiotic stages are rapidly accomplished after fertilization. This was observed *in vitro* by the appearance of polar bodies after fertilization (Beninger and Le Pennec, 2006).

The oogenesis in cephalopods is characterized by the invasion of the acinar ephitelium (named follicular ephitelium) into the oocyte and an extensive folding (Bottke, 1974). The lamellar sheets extend into the interior of the oocyte producing an amplification in the region of contact between accessory cells and oocyte surface membranes (Laptikhovsky and Arkhipkin, 2001).

# Stages of Oogenesis

## Premeiotic Stage

The premeiotic stage follows immediately upon oocyte formation. Oocytes could measures between 3.5-10 μm in bivalves (De Gaulejac et al., 1995; Eckelbarger and Davies, 1996; Chung, 2008) to 10-30 μm in gastropods (Arrighetti and Penchaszadeh, 2010). In some polyplacophora, as S. septentrionis, oogonia are located beneath the ovarian epithelium and as these differentiate into oocytes there is a corresponding growth of the epithelium. Some of the epithelial cells surround the oocytes and become accessory cells whereas others grow and become folds between differentiating oocytes (Selwood, 1968). In bivalves oogonia possess a large nucleus, in which chromatin is reticular an marginal. In Pinna nobilis the oogonia are maintained in close contact by desmosomes, allowing synchronous development (De Gaujelac et al., 1995). A similar morphology was observed in the snail *Adelomelon beckii* (Arrighetti and Penchaszadeh, 2010).

## Previtellogenesis

In this phase the nucleus progressively modificates and enlarges while the cytoplasm grows slowly as it synthesized several organelles, suggesting major synthetic activity.

In the ooplasm of previtellogenic oocytes (also called stage II oocytes) of the polyplacophora *Mopalia mucosa* and *Chaetopleura apiculata* are some pigment inclusions and lipids. A distinguishing feature of this stage is the presence of some acid mucopolysaccharide vacuolar areas, which sometimes occupy a large area within the oocyte cytoplasm (Andreson, 1972). In mature oocytes this area are found primarily within the peripheral cytoplasm and because of their position they were called cortical granules. At the follicle-oocytes interface appears the vitelline envelope.

In several bivalves the nucleus forms a nuclear ring in contact with the nuclear envelope (Pipe, 1987b; Dorange et al., 1989; De Gaujelac et al., 1995). In *Crassostrea virginica* previtellogenic oocytes are initially devoid of organelles, except for numerous perinuclear mitochondria and lipid doplets (Eckelbarger and Davies, 1996). Pipe (1987b) found evidence for uptake of macromolecules by pinocytosis in the basal region of previtellogenic oocytes of *M. edulis*.

## Vitellogenesis

Vitellogenic stage is characterized by a highly increase in the volume of the cytoplasm due to the storage of yolk and cell organelles. The vitellogenic stage is highly variable since the composition and organization of yolk differs among species and since different types of specialized cells may participate in the synthesis and transport of yolk or yolk precursors. The term yolk is used to mention a population of cytoplasmatic inclusions histochemically different: carbohydrates, lipids and proteins. The quality of an oocyte is measured by the amount of protein, lipids and carbohydrates present into the oocyte during yolk formation (Angel- Dapa et al., 2010). In particular, lipids have been considered an adequate indicator of oocyte quality in species such as Atrina maura (Angel-Dapa et al., 2010), Crassostrea corteziensis (Rodriguez-Jaramillo et al., 2008), Pinctada mazatlanica (Goméz-Robles et al., 2005).

Yolk protein is a major component of the cytoplasm of many oocytes and could be present in the form of droplets or platelets. In many molluscs, yolk protein is synthesized by the oocytes through autosynthetic process (Jong-Brink et al., 1983; Wourms, 1987; Eckelbarger and Davies, 1996). Nevertheless, the incorporation of exogenous yolk precursors by endocytosis is also well documented (Hill and Bowen, 1976; West, 1981; Wourms, 1987; Hodgson and Eckelbarger, 2000; Arrighetti and Penchaszadeh, 2010). The presence of lipids yolk has not been intensively study as that of protein yolk and it may occurs in the form of droplets or spheres.

In *M. mucosa* and *C. apiculata* carbohydrate-protein yolk bodies are initially seen as dense material within the saccules of the Golgi complex. At this time there is an increase in the amount of the rough endoplasmic reticulum. It is documented that the carbohydrate and protein precursors of the yolk bodies are synthesized by the endoplasmic reticulum and finally concentrated and packaged by the Golgi complex (Anderson, 1972). In mature oocytes yolk bodies are membrane-limited, bipartite, oval structures with the central region composed of a dense homogeneous component and a cortical region composed of a substance aligned as striations orientated perpendicular to the surface of the central component (Anderson, 1972). In the chiton *S. septentriones* protein yolk deposition is associated with the interstices formed at the onset of vitellogenesis. This increase in surface area of the plasma membrane exposed to the blood in the sinus enables the incorporation of protein from material in the blood stream (Selwood, 1968). In the case of lipid yolk, droplets first appear in the cytoplasm of the stalk region, indicating that the blood in the sinus is the source of lipids (Selwood, 1968).

Yolk synthesis in many bivalves takes place using both auto- and heterosynthetic pathways. Eckelbarger and Davies (1996) suggest for *C. virginica* that some yolk bodies are formed by the fusion of Golgi-derives vesicles followed by Golgi complex interaction with RER cisternae within the oocyte and some other by endocytotic activity. In the later, yolk precursors are transferred from the VCT cells to the oocytes of *C. virginica* using a hemocoelic pathway. Something similar occurs in females of *Patinopecten yessoensis* (Chung et al., 2005), *Meretaix lusoria* (Chung, 2007) and *Chlamys (Azumapecten) farreri farreri* (Chung, 2008) in were autosynthesis involved the combined activity of the Golgi complex, mitochondria, and RER, whereas heterosynthesis involved endocytotic incorporation of extraovarian precursors derived from accessory cells. Pipe (1987b) reported, for *M. edulis,* evidences for uptake of exogenous macromolecules into oocytes by pinocytosis during early stages of development. During winter months, lipid yolk development are in close association with mitochondria and glycogen-rich vesicles, therefore oocyte development takes place at expenses of stored reserves, mainly glycogen from VCT cells. However, during later vitellogenesis, yolk formation results from the activity of Golgi bodies and RER in the oocyte (Pipe, 1987b).

In some other bivalves yolk granules are formed using only autosynthetic pathway in where various cell organelles, in particular Golgi complex, RER, and mitochondria are thought to be involved in endogenous formation of yolk granules. De Gaulejac et al. (1995) state for *P. nobilis* that proteinaceous yolk granules are formed by fusion of vesicles derived from Golgi complex while lipid yolk granules are formed by coalescence of lipid globules and dilated cisternae of RER. Chung et al. published several papers reporting that Golgi complex, well-developed endoplamic reticulum, and mitochondria were involved in the formation of lipid droplets in the early vitellogenic oocytes of bivalves (Chung and Ryou, 2000; Chung et al., 2005; Chung, 2007; Chung, 2008).

Along vitellogenesis of the limpet *Siphonaria capensis* the increase in the number of RER and Golgi bodies indicates that the membrane-bound yolk granules are formed by autosynthesis. However, in *S. serrata* yolk granules are formed autosynthetically at the beginning of vitellogenesis and as vitellogenesis proceeds yolk appears to be formed heterosynthetically with coated pits developing along the oolema (Pal and Hodgson, 2002). Another example of yolk granules produced by both auto- and heterosynthesis is *A. beckii* (Arrighetti and Penchaszadeh, 2010). Exclusive autosynthetic yolk formation in gastropods were reported for *Neptunea (Barbitonia) arthritica cumingii* (Chung et al., 2006) and *Haliotis varia* (Najmudeen, 2008), although in *H. varia* there is a significant negative correlation between lipids and carbohydrates of the digestive gland and ovary, indicating the transfer of this biochemical components from the digestive gland to the ovary (Najmudeen, 2007).

In oocyte of the cephalopod *Octopus* probably all of the proteinaceous yolk is synthesized by the accessory cells and transferred to the oocyte via the characteristic infolding of the follicular epithelium (Wourms, 1987).

## Oocyte Degeneration

The degeneration and resorption of oocytes is a commonly observed phenomenon in molluscs (Jong-Brink et al., 1983; Wourms, 1987) which enable the recycling of nutrients to meet the energy requirements of basal metabolism.

The presence of big autophagic vacuoles and myelin-like organelles in accessory cells of several bivalves, suggest a major role in degradation and resorption of oocytes (Dorange and Le Pennec, 1989; De Gaulejac et al., 1995; Chung, 2008). In *M. edulis* oocyte degeneration involves initial breakdown of the plasma membrane followed by the rupture of the vitelline coat. The oocyte content once released into the acinar lumen is resorbed by lysosomal enzymes present in the epithelial cells and also in the yolk granules of degenerating oocytes (Pipe, 1987b). In the case of *C. farreri farreri,* during the period of oocyte degeneration, the function of accessory cells are phagocytosis and intracellular digestion of products originating from oocyte degeneration (Chung, 2008), the same occurs in *Meretaix lusoria* (Chung, 2007). During this period accessory cells probably have a lysosomal system for breakdown, and resorb phagosomes in the cytoplasm for nutrient storage (Chung, 2007, 2008). In *A. beckii*, Arrighetti and Penchaszadeh (2010) reported considerable morphological evidences for phagocytosis of material by accessory cells, with a large amounts of RER and a well-developed lysosomal system.

## Oocytes Envelopes

In mollusc three types of oocyte envelopes could be distinguished: primary envelopes are formed within the ovary, secondary envelopes are produced by accessory cells and tertiary envelopes formed by the accessory sex glands.

Most of the molluscan oocytes have a primary egg envelope (also named vitelline envelope), at least at some stage during their differentiation. Sometimes this primary egg envelope is a bipartite or tripartite structure and acts as fertilization membranes (also called "jelly coats"). Oocytes of some bivalves, as *Mytilus edulis*, *Pinna nobilis*, possess a primary egg envelope formed by an homogeneous filamentous protein-polysaccharide complex. The cortical granules present in the periphery of their oocytes slowly release mucus to the perivitelline coat. Spawned eggs in some species of the abalone *Haliotis* are surrounded by a primary egg envelope that contain three distinct elements: the jelly layer, the vitelline envelope, and the egg surface coat. The sperm pass through the jelly coat and when binds the vitelline envelope appears to trigger the acrosome reaction. Then, the lysine released form the acrosome dissolves a hole in the vitelline envelope through which sperm swims (Mozingo et al., 1995).

Secondary egg envelopes are synthesized and secrete by accessory cells. Polyplacophorans are known to have secondary eggs envelopes that are known as egg hulls. Polyplacophorans of the order Chitonida, with the exception of *Callochiton dentatus* (Buckland-Nicks and Reunov, 2010), have egg hulls ornamented with elaborated extra-cellular coverings like cupules, cups, cones, flaps or spines (Buckland-Nicks, 2008; Ituarte et al., 2010), while members of the basal order Lepidopleurida has eggs with smooth jelly-like hulls (Buckland-Nicks and Hogdson, 2000; Buckland-Nicks, 2008). This variation in the morphology of the egg hull projections is a valuable character in the study of phylogenetic affinities (Buckland-Nicks, 2008; Ituarte et al., 2010). The mechanism of egg hull formation in females of *C. dentatus* was intensively studied by Buckland-Nicks and Reunov (2010) and concluded that is quite different from that reported for members of other species of chitons (Selwood, 1968; Anderson, 1972; Buckland-Nicks and Renuov, 2009). In *C. dentatus* microapocrine secretions released by the oocyte are the primarily responsible for egg hull formation. Accessory cells are responsible for gather the mixture of oocyte secretions and draw them out into radial stripes while in members of some other families in the order Chitonida form their egg hulls with secretory inputs from the accessory cells (Buckland-Nicks and Reunov, 2010).

The other major group in where secondary egg envelope is present is cephalopods. In this group the egg is surrounded by a rigid secondary envelope, named "chorion", derived from the accessory cells. This impenetrable barrier can only be traversed by sperm through a passageway termed micropyle, as was studied in some species of the order Theuthida and Octopoda (Wourms, 1987).

The tertiary egg envelopes are formed after the egg has left the ovary and are present in certain bivalves, chitons and most gastropods, especially in caenogastropods. This envelope is secreted by cells in the oviduct or by accessory sex glands and is present in organisms that deposit their eggs on a substrate or brood them (Wourms, 1987). In gastropods, the tertiary egg envelops serve for protection and nutrition. The eggs are surrounded with albumen, protein and polysaccharides secreted by the albumen gland and later enclosed either in gelatinous masses or egg capsules formed by the capsule gland. A detailed description was made by Fretter and Graham (1994).

The tertiary egg envelope of some cephalopods is formed by oviductal glands, nidamental glands, and accessory nidamental glands which produces a protein-contain material that is added to the secondary egg envelope (chorion) and serves as an adhesive to the substrate (Wourms, 1987).

## Oogenesis and Pollution

In recent years, levels of contaminants in the marine environment have increased as a consequence of anthropogenic activities (Cajaraville et al., 2000). In some species oocyte degeneration and resorption may result from exposure to environmental contamination by pollutants.

Organotin compounds are ubiquitous contaminants in the environment and their effects were intensively studied in molluscs. In particular tributyltin (TBT), which is used mainly as a biocide in various boat antifouling paints and wood preservatives, is one of the most potent androgenic compounds ever added to the environment. TBT causes the imposition of masculine characters over female gastropods and this phenomenon, named imposex, is known to occur in more than 200 species of marine gastropods (Bigatti et al., 2009 and references therein). The effects of TBT depends on the species, in some cases does not impair reproduction (Amor et al., 2004) while in some others it causes the imposex female to have reproductive failure and finally leads to a population decline (Oehlmann et al., 1996; Horiguchi et al., 2008). At histological level, in some species, TBT could induce spermatogenesis in the ovaries of imposex females (Horiguchi et al., 2002, 2006) and/or have negatives effects on oogenesis (Gibbs et al., 1988; Oehlmann et al., 1996; Horiguchi et al., 2002)

Imposex females of *Babylonia areolata* demonstrate oocyte degeneration as indicated by the presence of numerous lipids droplets in the ovarian tissues (Muenpo et al., 2010). In the gastropod *Nucella lapillus* as a consequence of TBT exposure in early life stages a sex change might occur and oogenesis is suppressed and supplanted by spermatogenesis (Oehlmann et al., 1991). Other endocrine disruptor that are known to have negative effects on marine and freshwater snails are bisphenol A (BPA) and octylphenol (OP). Oehlmann et al. (2000) studied the effect of both chemicals on females of *Marisa cornuarietis* and concluded that the affected specimens were characterized by a massive stimulation of oocyte and spawning mass production.

## REFERENCES


Amor, M. J., Ramón, M., & Durfort, M. (2004). Ultrastructure studies of oogenesis in *Bolinus brandaris* (Gastropoda: Muricidae). *Scientia Marina,* 68, 343-353.

Angel-Dapa, M. A., Rodriguez-Jaramillo, C., Cáceres-Martínez, C. J. & Saucedo, P. E. (2010). Changes in lipid content of oocytes of the penshel Atrina maura as a criterion of gamete development and quality: A study of histochemistry and digital image analysis. *Journal of Shellfish Research,* 29, 407-413.

Anderson, E. (1972). Oocyte-follicle cell differentiation in two species of amphineurans (Mollusca), Mopalia mucosa and Chaetopleura apiculata, In: G. Taylor (Ed.), *Spiralian Embryology.* (1st edition, pp. 7-43) Ardent Media.

Arrighetti, F. & Penchaszadeh, P. E. (2010). Gametogenesis, seasonal reproduction and imposex of *Adelomelon beckii* (Neogastropoda: Volutidae) in Mar del Plata, Argentina. *Aquatic Biology,* 9, 63-75.

Beninger, P. G., Le Pennec, G., Le Pennec, M. (2003). Demonstration of nutrient pathway from the digestive system to oocytes in the gonad intestinal loop of the scallop Pecten maximus L. *Biological Bulletin,* 205, 83-92.

Beninger, P. G. & Le Pennec, M. (2006). Structure and function in scallops. In: S. E. Sumway & G. J. Parson (Eds.), *Scallops: biology, ecology and aquaculture.* (1st edition, 1460pp. 123-228) Amsterdam: Elsevier B.V.

Bigatti, G., Primost, M. A., Cledón, M., Averbuj, A., Theobald, N., Gerwinski, W., Arntz, W., Morriconi, E. & Penchaszadeh, P. E. (2009). Biomonitoring of TBT contamination and imposex incidence along 4700 km of Argentinean shoreline (SW Atlantic: From 38S to 54S). *Marine Pollution Bulletin,* 58, 695-701.

Bottke, W. (1974) The fine structure of the ovarian follicle of *Alloteuthis subulata* Lam. (Mollusca, Cephalopoda). *Cell and Tissue Research,* 150, 463-479.

Bottke, W., Sinha, I. & Keil, I. (1982) Coated vesicle-mediated transport and deposition of vitellogenic ferritin in the rapid growth phase of snail oocytes. *Journal of Cell Science,* 53, 173-191.

Brusca, R. C. & Brusca, G. J. (2002). *Invertebrates* (2nd edition). Massachussets: Sinauer Associates, Inc.

Buckland-Nicks, J., & Hodgson, A. N. (2000). Fertilization in Callochiton castaneus (Mollusca). *Biological Bulletin,* 199, 59–67.

Buckland-Nicks, J. (2008) Fertilization biology and the evolution of chitons. *American Malacological Bulletin,* 25, 97-111.

Buckland-Nicks, J. & Reunov, A. (2009). Ultrastructure of hull formation during oogenesis of *Rhyssoplax tulipa* (=Chiton tulipa) (Chitonidae: Chitoninae). *Invertebrate Reproduction and Development,* 53, 165–174.

Buckland-Nicks, J. & Reunov, A. A. (2010). Egg hull formation in *Callochiton dentatus* (Mollusca, Polyplacophora): The contribution of microapocrine secretion. *Invertebrate Biology,* 129, 319-327.

Cajaraville, M. P., Bebianno, M. J., Blasco, J., Porte, C., Sarasquete, C. & Viarengo, A., (2000). The use of biomarkers to assess the impact of pollution in coastal environments of the Iberian Peninsula: a practical approach. *The Science of the Total Environment,* 247, 295-311.

Chung, E. Y., Park, Y. J., Lee, J. Y. & Ryu, D. K. (2005). Germ cell differentiation and sexual maturation of the hanging cultured female scallop Patinopecten yessoensis on the east coast of Korea. *Journal of Shellfish Research,* 24, 913-921.

Chung, E. Y., Kim, S. Y., Park, G. M. & Yoon, J. M. (2006). Germ cell differentiation and sexual maturation of the female *Neptunea* (Barbitonia) *arthritica cumingii* (Crosse, 1862) (Gastropoda: Buccinidae). *Malacologia,* 48, 65−76.

Chung, E. Y. (2007). Oogenesis and sexual maturation in *Meretrix lusoria* (Röding, 1798) (Bivalvia: Veneridae) in western Korea. *Journal of Shellfish Research,* 26, 1−10.

Chung, E. Y. (2008). Ultrastructural studies of oogenesis and sexual maturation in female *Chlamys* (Azumapecten) *farreri farreri* (Jones & Preston, 1904) (Pteriomorphia: Pectinidae) on the western coast of Korea. *Malacologia,* 50, 279-292.

Chung, E. Y. & Ryou, D. K. (2000). Gametogenesis and sexual maturation of the surf clam *Mactra veneriformis* on the west coast of Korea. *Malacologia,* 42, 149–163.82

Currie, D. R. (1990). Reproductive periodicity of three species of chitons at a site in New South Wales, Australia (Mollusca: Polyplacophora). *Invertebrate Reproduction and Development,* 17, 25-32.

De Gaulejac, B., Henry, M. & Vicente, N. (1995). An ultrastructural study of gametogenesis of the marine bivalve *Pinna nobilis* (Linnaeus 1758) I. Oogenesis. *Journal of Molluscan Studies,* 61, 375-392.

Dorange, G. & Le Pennec, M. (1989). Ultrastructural study of oogenesis and oocytic degeneration in *Pecten maximus* from the Bay of St. Brieuc. *Marine Biology,* 103, 339-348.

Eckelbarger, K. J. (1994). Diversity of metazoan ovaries and vitellogenic mechanisms: implications for life history theory. *Proceeding of the Biological Society of Washington,* 107, 193-218.

Eckelbarger, K. J. & Davies, C. V. (1996). Ultrastructure of the gonad and gametogenesis in the eastern oyster, *Crassostrea virginica*. I. Ovary and oogenesis. *Marine Biology,* 127, 79-87.

Fretter, V. & Graham, A. (1994). *British prosobranch molluscs: their functional anatomy and ecology.* (2nd edition). London: The Ray Society.

Giese, A. C. & Pearse, J. S. (1974). Introduction: General Principles. In A. C. Giese & J. S. Pearse (Eds.) *Reproduction of Marine Invertebrates,* Volume I (1st edition, pp. 1-49). New York, USA: Academic Press.

Giménez, J. & Penchaszadeh, P. E. (2002). Reproductive cycle of *Zidona dufresnei* (Caenogastropoda: Volutidae) from the southwestern Atlantic Ocean. *Marine Biology*, 140, 755-761.

Gibbs , P.E., Pascoe, P.L. & Burt, G.R. (1988). Sex change in the female dogwhelk, *Nucella lapillus*, induced by tributyltin from antifouling paints. *Journal of the Marine Biology Association of the United Kingdom,* 68, 715–731.

Gómez-Robles, E., Rodriguez-Jaramillo, C. & Saucedo, P. E. (2005). Digital image analysis of lipid and protein histochemical markers for measuring oocyte development and quality in pearl oyster *Pinctada mazatlanica* (Hanley, 1856) *Journal of Shellfish Research,* 24, 1197-1202.

Griffond, B. & Gomot, L. (1979). Ultrastructural study of the follicle cells in the freshwater gastropod *Viviparus viviparus* L. *Cell and Tissue Research,* 202, 25-32.

Hayashi, I. (1980). The reproductive biology of the ormer, *Haliotis tuberculata. Journal of the Marine Biology Association of the United Kingdom,* 60, 415-430.

Hickman, C. P., Roberts, L. S. & Larson, A. (2001*). Integrated principles of zoology (11th edition).* New York: McGraw-Hill.

Hill, R. S. & Bowen, I. D. (1976). Studies on the ovotestis of the slug Agriolimax reticulatus (Müller). *Cell and Tissue Research,* 173, 465-482.

Hodgson, A. N. & Eckelbarger, K. J. (2000) Ultrastructure of the ovary and oogenesis in six species of patellid limpets (Gastropoda: Patellogastropoda) from South Africa. *Invertebrate Biology,* 119, 265-277.

Horiguchi, T., Kojima, M., Kaya, M., Matsuo, T., Shiraishi, H., Morita, M. & Adachi, Y. (2002). Tributyltin and triphenyltin induce spermatogenesis in ovary of female abalone, *Haliotis gigantea. Marine Environmental Research,* 54, 679-684.

Horiguchi, T., Kojima, M., Hamada, F., Kajikawa, A., Shiraishi, H., Morita, M. & Shimizu, M. (2006). Impact of tributyltin and triphenyltin on ivory shell (Babylonia japonica) populations. *Environmental Health Perspectives,* 114, 13-19.

Ilano, A. S., Fujinaga, K. & Nakao, S. (2003). Reproductive cycle and size at sexual maturity of the commercial whelk Buccinum isaotakii in Funka Bay, Hokkaido, Japan. *Journal of the Marine Biology Association of the United Kingdom,* 83, 1287-1294.

Ituarte, C., Liuzzi, M.G. & Centurión, R. (2010). Egg hull morphology in two chitons (Polyplacophora) from the Southwestern Atlantic Ocean. *Malacologia,* 53, 167–174.

Jong-Brink, M., Boer, H. H. & Joose, J. (1983). Mollusca. In K. G. Adiyodi & R. G. Adiyodi (Eds.) Reproductive Biology of invertebrates. *Volume 1: Oogenesis, oviposition and oosoption* (1st edition, pp. 297-355) John Wiley & Sons Ltd.

Laptikhovsky, V. V. & Arkhipkin, A. I. (2001). Oogenesis and gonad development in the cold water loliginid squid *Loligo gahi* (Cephalopoda: Myopsida) on the Falkland Shelf. *Journal of Molluscan Studies,* 67, 475-482.

Le Pennec, M., Dorange, G., Beninger, P. G., Donval, A.& Widowati, I. (1991). Les relations trophiques anse intestinale-gonade chez *Pecten maximus* (Mollusque, Bivalve). *Haliotis,* 21, 57–69.

Lowe, D. M., Moore, M. N. & Bayne, B. L. (1982). Aspects of gametogenesis in the marine mussel *Mytilus edulis* L. *Journal of the Marine Biology Association of the United Kingdom,* 62, 133-145.

Martel, A., Larrivée, D. H., Klein, K. R. & Himmelman, J. R. (1986). Reproductive cycle and seasonal feeding activity of the neogastropod *Buccinum undatum*. *Marine Biology,* 92, 211-221.

Mozingo, N. M., Vacquier, V. D. & Chandler, D. E. (1995) Structural features of the abalone egg extracellular matrix and its role in gamete interaction during fertilization. *Molecular Reproduction and Development,* 41, 493–502.

Muenpo, C., Suwanjarat, J. & Klepal, W. (2010). Ultrastructure of oogenesis in imposex females of *Babylonia areolata* (Caenogastropoda: Buccinidae). *Helgoland Marine Research,* 65, 335-345.

Nagabhushanam, R. & Deshpande, V. D. (1982). Reproductive cycle of the chiton *Chiton iatricus* and environmental control of its gonad growth. *Marine Biology,* 67, 9-15.

Najmudeen, T. M. (2007). Variation in biochemical composition during gonad maturation of the tropical abalone *Haliotis varia* Linnaeus, 1758 (Vetigastropoda: Haliotidae). *Marine Biology Research,* 3, 454–461.

Najmudeen, T. M. (2008). Ultrastructural studies on the ovary maturation of the variable abalone *Haliotis varia* Linnaeus 1758 (Vetigastropoda: Haliotidae). *Aquatic Biology,* 2, 143-151.

Oehlmann, J., Stroben, E. & Fioroni, P. (1991). The morphological expression of imposex in *Nucella lapillus* (Linnaeus) (Gastropoda: Muricidae). *Journal of Molluscan Studies,* 57, 375-390.

Oehlmann, J., Fioroni, P., Stroben, E. & Markert, B. (1996). Tributyltin (TBT) effects on *Ocinebrina aciculata* (Gastropoda: Muricidae): Imposex development, sterilization, sex change and population decline. *Science of the Total Environment,* 188, 205–223.

Oehlmann, J., Schulte-Oehlmann, U., Tillmann, M. & Markert, B. (2000). Effects of endocrine disruptors on prosobranch snails (Mollusca: Gastropoda) in the laboratory. Part I: bisphenol A and octylphenol as xeno-estrogens. *Ecotoxicology,* 9, 383-397.

Ortiz, N., Ré, M. E., Márquez, F. & Glembocki, N. G. (2011). The reproductive cycle of the red octopus *Enteroctopus megalocyathus* in fishing areas of Northern Patagonian coast. *Fisheries Research,* 110, 217-223.

Pal, P. & Hodgson, A. N. (2002). An ultrustructural study of oogenesis in a planktonic and a direct-developing species of siphonaria (Gastropoda: Pulmonata). *Journal of Molluscan Studies,* 68, 337-344.

Pipe, R. K. (1987a). Ultrastructural and cytochemical study on interactions between nutrient storage cells and gametogenesis in the mussel *Mytilus edulis*. *Marine Biology,* 96, 519-528.

Pipe, R. K. (1987b). Oogenesis in the marine mussel *Mytilus edulis*: an ultrastructural study. *Marine Biology,* 95, 405-414.

Rodriguez-Jaramillo, C., Hurtado, M. A., Romero-Vivas, E., Ramírez, J. L., Manzano, M. & Palacios, E. (2008). Gonadal development and histochemistry of the tropical oyster, *Crassostrea corteziensis* (Hertlein, 1951) during an annual reproductive cycle. *Journal of Shellfish Research,* 27, 1129-1141.

Rodríguez-Rúa, A., Pozuelo, I., Prado, M. A., Gómez, M. J. & Bruzón, M. A. (2005). The gametogenic cycle of *Octopus vulgaris* (Mollusca: Cephalopoda) as observed on the Atlantic coast of Andalusia (south of Spain). *Marine Biology,* 147, 927-933.

Rosa, R., Costa, P. R. & Nunes, M. L. (2004). Effect of sexual maturation on the tissue biochemical composition of *Octopus vulgaris* and *O. defilippi* (Mollusca: Cephalopoda). *Marine Biology,* 145, 563-574.

Ruppert, E. E. & Barnes, R. D. (1995). *Zoología de los invertebrados* (6th edition). Mexico, D.F.: McGraw-Hill Interamericana.

Schroeder, P. C. (1987). Endogenous control of gametogenesis. In A. C. Giese & J. S. Pearse (Eds.), Reproduction of marine invertebrates. Volume IX. *General aspects: Seeking unit in* diversity (1st edition, pp. 179-249) New York, USA: Academic Press.

Selwood, L. (1968). Interrelationships between developing oocytes and ovarian tissues in the chiton *Sypharochiton septentriones* (Ashby) (Molluscar, Polyplacophora). *Journal of Morphology,* 125, 71-104.

Selwood, L. (1970). The role of the follicle cells during oogenesis in the chiton *Sypharochiton septentriones* (Ashby) (Polyplacaphora, Mollusca) *Cell and Tissue Research,* 104, 178-192

Son, P. W. & Chung, E. Y. (2009). Annual reproductive cycle and size at sexual maturity of the sun and moon scallop *Amusium japonicum japonicum* (Gmelin, 1791) (Bivalvia: Pectinidae) in the coastal waters of Jejudo, Korea. *Malacologia,* 51, 119-129.

West, D.L. (1981). Reproductive biology of *Colus stimpsoni* (Prosobranchia: Buccinidae). IV. Oogenesis. *The Veliger,* 24, 28-38.

Wourms, J. P. (1987). Oogenesis. In A. C. Giese & J. S. Pearse (Eds.), Reproduction of marine invertebrates. Volume IX. *General aspects: Seeking unit in diversity* (1st edition, pp. 50-179) New York, USA: Academic Press.

In: Spawning
Editor: Erick R. Baqueiro Cárdenas

ISBN: 978-1-63117-655-5


*Chapter 3*

# SPERMATOGENESIS AND SPERM MORPHOLOGY

***Juliana Giménez***
Biología de Invertebrados Marinos. DBBE. Facultad de Ciencias Exactas y Naturales. Universidad de Buenos Aires, Argentina
IBBEA, CONICET- UBA

## ABSTRACT

This study describes the revision of spermatogenesis and paraspermatogenesis process in gastropods and comparative sperm morphology in the most of the classes of Molluscs.

The spermatogenesis pattern including, the nuclear condensation involved granular, fibrillar, lamellar, and final homogenous electron-dense phases. Acrosome development starts with the posteriorly-located proacrosomal vesicle arising from the Golgi complex in euspermatids. This proacrosomal vesicle develops into a pre-attachment acrosome, which, together with the Golgi body, later moves towards the apex of the nucleus.

Ultrastructural studies show that spermatozoa are useful indicators of systematic position and phylogenetical relationship within the Molluscs.

The morphology of the mature sperm of Bivalves, Solenogasters, Monoplacophora, Cephalopoda and Gastropoda, given in this work, can

* Intendente Güiraldes 2160, 4 ° piso. Ciudad Universitaria - C1428EGA . Buenos Aires. Argentina.

be used to understand some aspect of the reproduction, how the shape of the sperm can be related to internal or external fertilization.

Also is describes the dimorphic types of sperms in gastropods, call parasperm, and this related to the mode and the capacity of fertilization and the morphology of the parasperm also can be used as indicators of systematic position.

Paraspermatogenesis is describes in gastropods and the pattern of chromatin distribution in dense patches, together with the cytoplasmatic characteristics of paraspermatogonia, allow for the recognition of the apyrene line from the euspermatogonia. Later-occurring features, including the peripheral condensation of nuclear chromatin followed by nuclear invagination, the posterior breakdown into nuclear vesicles ("caryomerites"), centriole multiplication, and the synthesis of secretory products, are the most conspicuous changes in the paraspermatogenesis process.

## Introduction

The diversity in Mollusca is comprised of many classes, subclasses, orders, and species. Molluscs have been very successful in their main habitats, and all of them are represented in marine habitats.

Invertebrate spermatozoa are divided in three main types: aquasperm (ectaquasperm and entaquasperm), introsperm and dimorphic or polymorphic sperm (Jamieson, 1987; Rouse and Jamieson, 1987). Ectaquasperm are primitive and shed into the water. This 'primitive' sperm type is characterized by a small head, containing a rounded or conical nucleus, surmounted by a cup-shaped acrosomal vesicle, a middle piece with a ring of 4–5 rounded mitochondria encircling two centrioles, and a tail flagellum about 50 μm long (axoneme with a 9+2 microtubule pattern) (Franzen, 1983).

One of the most studied groups in Gastropoda are neogastropods where the presence of sperm dimorphism is pronounced. Ultrastructural studies show that sperm morphology is a useful indicator of systematics position and phylogenetical relationship within the Gastropoda (Healy, 1988a)

## Spermatogenesis

The ultrastructural characteristics of spermatogenesis revealed important findings showing that different groups of organisms present specific morphogenetic traits, including: (1) the cell stage at which proacrosomal vesicles first appear; (2) the mechanism by which the acrosomal vesicle migrates towards the apical pole of the cell; (3) the morphological appearance and chemical constitution of the acrosome subcomponents; (4) the origin of the subacrosomal material; (5) the pattern of chromatin remodelling during condensation; (6) and the mitochondrial composition and structure of the middle piece (Sousa and Azevedo, 1988; Sousa et al., 1989 ). The comparative study of sperm development is useful in mollusc phylogenetic studies. Here we present characteristics in the Classes Gastropoda that are well represented in the literature.

## GASTROPODA

Within caenogatropods the process of spermatogenesis can be distinguished as euspermatogenesis when describing the development of typical sperms (eusperm and paraspem) known as euspermatozoa. Zabala et al. (2012) described the processes of euspermatogenesis in the caenogastropod *Adelomelon ancilla*. The process occurs within every spermatogenic tubule (Figure 1, 2). Euspermatogenic cells are generally clustered in small groups at the same maturation phase and connected by a cytoplasmatic continuity. These groups of cells are distributed throughout the tubule (Figures 1, 2, 4). Once the process of euspermiogenesis is complete, the mature euspermatozoa fills the lumen of the tubule. Prior to the completion of euspermiogenesis, euspermatozoa also occurs scattered among euspermatogonia and euspermatocytes, attached to the Sertoli cells (Figure 3, inset).

The euspermatogonias are difficult to distinguish from early euspermatocytes. However, these cells possess an irregularly shaped nucleus, roughly spherical, and they generally occurred close to the periphery of spermatogenic tubules, lying just beneath the basal membrane (Figs. 1, 3).

Inset: Sertoli cells with spermatozoa attached. 4. Clusters of euspermatocytes connected by cytoplasmic bridges (arrowhead). Note advanced stages of euspermatids and bundle of mature sperm. Note also the Sertoli cell nucleus ovoid. Abbreviations: bm, basal membrane; ca, caryomerites; f, flagella; pg, paraspermatogonia; pt, paraspermatid; pz, paraspermatozoa; sc, euspermatocytes; sg, euspermatogonia; st, euspermatid; sz, mature euspermatozoa; ser, Sertoli cell. Scale bars: 1–4 = 20 mm; 3 inset = 10 μm. (Modified from Zabala et al., 2012). Zabala et al. ( 2012) reported measurements for the euspermatogonia of *A. ancilla* with a diameter of 11–13 μm with a large, round nucleus (11 μm in diameter) occupying nearly the entire volume within the cell. Small clumps of electron-dense chromatin are loosely distributed throughout the nucleoplasm. The single and conspicuous nucleolus is homogeneous and eccentrically located (Figure 5) and the scanty cytoplasm contains rounded mitochondria (Figure 5).

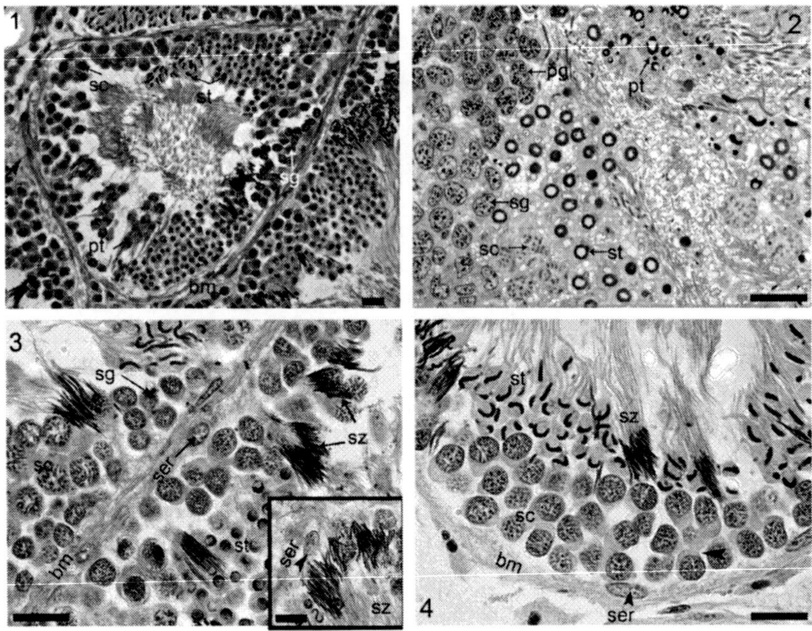


Figures 1-4. Light microscope sections of the testis of *Adelomelon ancilla.* 1, 3–4: Modified Masson's trichrome technique. 1. Transverse section of a spermatogenic tubule surrounded by connective tissue (arrowhead). Note both spermatozoa types in same tubule. 2. Detail of different spermatogenic stages. Semithin section stained with toluidine blue. 3. Detail of a cluster of euspermatogonia lying adjacent to the basal membrane. Note different stages of euspermatocytes and developmental stages of euspermatids. A bundle of mature euspermatozoa is visible attached to the Sertoli cell.

Euspermatocytes are 13–14 mm in diameter; presenting a small amount of cytoplasm surrounding a nucleus that measures 10–12 mm in diameter. The chromatin is typically more uniformly dispersed throughout the nucleoplasm (Figure 6). Synaptonemal complexes are frequently observed during this stage (Figure 6). First and second meiotic divisions apparently occur in rapid succession, and secondary euspermatocytes are rarely seen. The secondary euspermatocytes shows conspicuous patchy clumps of chromatin, and their mitochondria are located in the proximity to the basal nucleus (Figure 6). In addition to the elements found in the euspermatogonia, the euspermatocyte cytoplasm contains several round mitochondria and a well-developed Golgi complex.

Euspermatids were also described in the work of Zabala et al. (2012). In order to resume the process of spermatogenesis Zabala et al. (2012) presented a diagrammatic step of the euspermatogenesis. (Figures 5-9). Euspermatids are formed in groups, interconnected by cytoplasmic bridges (Figure 7).

In general, they are located throughout the luminal region and possessed a smaller nucleus than the euspermatogonia and euspermatocyte. The late-stage euspermatocytes are distinguished from the early stages by the generally denser appearance of chromatin and the numerous organelles in the cytoplasm. The euspermatids undergo many changes in the shape of the nucleus, as well as remarkable changes in the fine structure of the cytoplasm. The euspermatid at the ring stage is roughly spherical (6 mm in diameter), with a central round nucleus (Figure 7).

It is characterized by the condensation of nuclear granular chromatin at the periphery, while the nucleoplasm of the central region preserve the original granular stage, forming a ring (Figure 7). The mitochondria distributed throughout the cytoplasm aggregates at the posterior pole of the nucleus. Acrosome formation starts with the pro-acrosomal vesicle arising from the Golgi complex, both located at the posterior pole of the cell (Figure 8).

The remaining vesicles coalesced with the largest vesicle to form the proacrosomal granule, which invaginates on its inner surface, forming a cup shape. Fibrillar euspermatid stage shows longitudinal arrangement of thickening fibrils of chromatin. The axoneme is inserted into the differentiated zone while a well-developed Golgi complex and mitochondria starts the elongation phase (Figure 9).

The centriolar complex is inserted into the nuclear fossa located at the 'differentiated zone'. The mitochondria starts the elongation phase around the axoneme. The almost complete preacrosome complex begins migration, together with Golgi complex, towards apex of the nucleus. At this stage cytoplasmic bridges are also visible (Figure 8). In the last stage of the spermatid, chromatin condensation acquires the form of lamellar plates, very elongated and twisted. Acrosome complex reaches final position at apex of nucleus while the Golgi complex migrates posteriorly, lateral to nucleus. A reduced endoplasmic reticulum is also visible in cytoplasm (Figure 9).

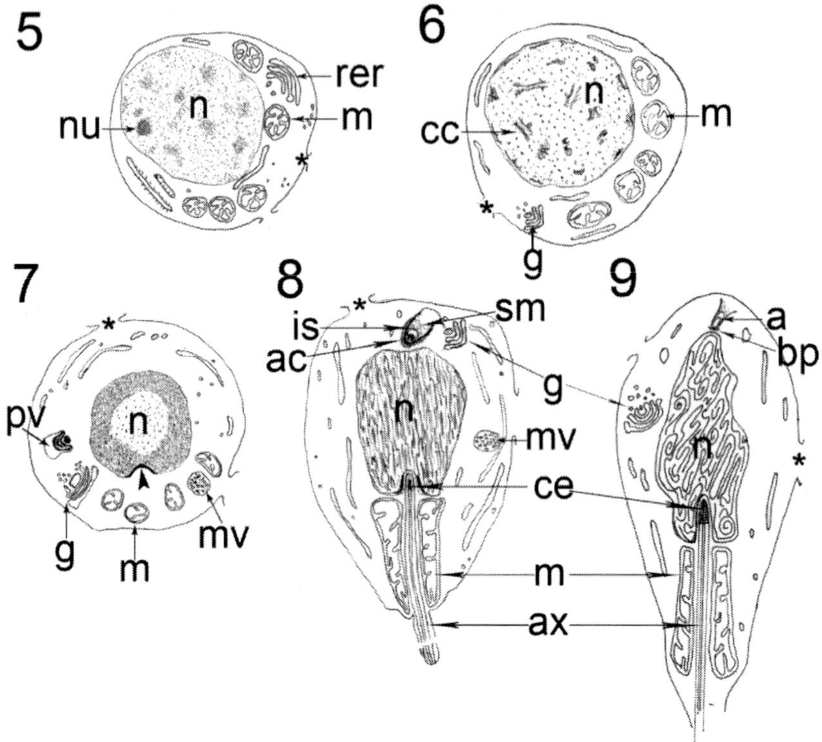


Figurs 5-9. Diagram of euspermatogenesis in *Adelomelon ancilla*. Not exactly to scale. 5. Euspermatogonium. Note chromatin pattern distributed throughout nucleoplasm. 6. Euspermatocyte. Note chromosome cores. 7. Ring euspermatid stage. Acrosome formation has begun close to Golgi complex, both located at the base of cell. 8. Fibrillar euspermatid stage. 9. Lamellar euspermatid stage. Abbreviations: a, acrosome; ac, acrosome cone; ax, axoneme; arrowhead, differentiated zone; asterisk, cytoplasmic bridges; bp, basal plate; cc, chromosome cores; ce, centriolar complex; g, Golgi complex; is, inner supporting structure; m, mitochondria; mv, multivesicular body; n, nucleus; nu, nucleolus; pv, proacrosomal vesicle; rer, rough endoplasmic reticulum; sm, subacrosomal material. (From Zabala et al., 2012).

## Sperm Morphology

The interpretation of the structure and function of the mature spermatozoa is often facilitated by a comparative study of spermatogenesis. The sperm found in many primitive aquatic organisms, is called the primitive type by Retzius (1904, 1905). The primitive sperm is a small cell with an ovoid head, with nucleus and acrosome and a short midpiece, arranged by 4 or 5 mitochondrias in the base of the nucleus. The tail is a long flagellum. This shape type is generally related to external fertilization.

## Acrosome

The acrosome is situated anterior to the nucleus; this structure enables the spermatozoon to penetrate the protective layers of the egg. The process from the acrosome depends from the activity of Golgi vesicles. It is possible to observe the vesicles and the well-developed Golgi complex, during the formation of the acrosomal complex (Figure 9).

The morphology of these structures varies in relation to functional demands during the fertilization process, and in relation to the taxonomical position of the species (Franzen, 1987).

## The Nucleus

The condensed chromatin containing the paternal hereditary material is transformed from spherical shape in spermatocytes and early spermatids to a compact body in mature spermatozoa. Generally the nucleus is the most electron-dense portion from the spermatozoon.

## The Middle Piece

The middle piece is composed by the mitochondrial material, and attains different shapes in molluscs spermatozoa.

Among bivalves, polyplacophorans and the more primitive gastropods, the spermatozoa have middle pieces with relatively unmodified mitochondria located at the base of the nucleus.

In snails as *Littorina* and *Nucella,* the mitochondria are scattered throughout the cytoplasm in early spermatid stages. During spermiogenesis, the mitochondria aggregate at the base of the nucleus and become larger and ovoid in shape (Buckland-Nicks and Chia, 1973).

In some Neogastropoda, the mitochondria become elongated during spermiogenesis, and finally become elongated and helically (Figure 9) arranged around the axoneme (Giménez el al, 2008, Giménez, 2011, Zabala et al., 2012)

In cephalopods, the middle piece present different complicated arrangements (Longo and Anderson, 1970, Maxwell, 1974), with varying number of mitochondrial columns surrounding the axoneme posterior to the nucleus (Figure 7)

## Glycogen Piece

The glycogen piece is localized surrounding the axoneme, posterior to the middle piece of the spermatozoa. In this piece, it is possible to note a reserve of stored glycogen in the sperm. The polysaccharide presumably serves as an endogenous source of energy.

## Bivalvia

The spermatozoa are shed to the surrounding media. In other molluscs where the spermatozoa are not discharged freely into the water, the morphology of the spermatozoon is modified.

The spermiogenesis brings new structures to contribute with the functional demands.

In bivalves, there is a tendency towards evolution of an elongated sperm nucleus in species with large eggs, whereas the middle piece retains its surrounding two centrioles (Franzen, 1983).

Within the Veneroidea (Bivalvia, Heterodonta), the spermatozoa are of the 'primitive' type in almost all species studied. Although, this definition includes some modifications of the sperm nucleus and middle piece that deviate from the most common type (Lucinacea: Healy, 1995; Johnson et al., 1996; Leptonacea: Galeommatidae: Eckelbarger et al., 1990; Cardiacea: Sousa & Azevedo, 1988; Healy, 1995; Keys and Healy, 1999, 2000; Tellinacea: Sousa et al., 1989; Healy, 1995; Dreissenacea: Denson and Wang, 1994, 1998; Walker et al., 1996; Veneracea: Turtoniidae: Ockelmann, 1964).

A few examples of modified and dimorphic sperm (Lucinacea: Healy, 1995; Moueza and Frankiel, 1995; Leptonacea: Lasaeidae: O'Foighil, 1985; Leptonacea: Montacutidae: Ockelmann, 1965; Corbiculacea: Komaru and Konishi, 1996; Konishi et al., 1998).

## Solenogasters

The mature spermatozoa possess an acrosome, a rod-shaped nucleus, and elongated filiform middle piece and end piece (Figure 10). (Franzen, 1955b; Buckland Nicks and Scheltema, 1995). These characteristics are also founded in the caenogastropods sperm.

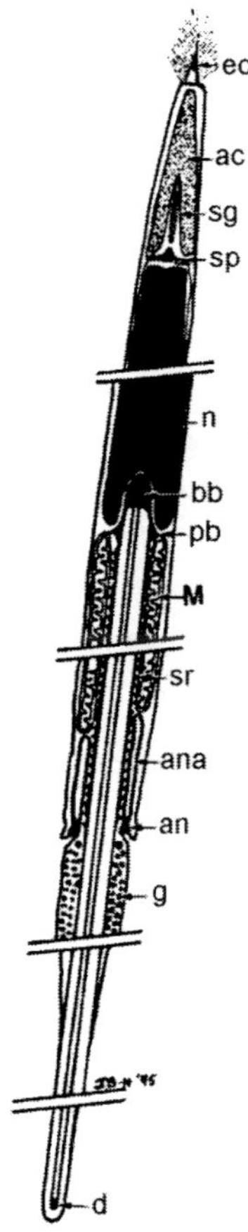


Figure 10. Spermatozoa of Solenogasters *Epimenia australis*. Abbreviation: ac, mature acrosomal; an, annulus ; ana,anular adjunt; bb, basal body (centriolar region); d, dense body and termination of axonemes; ec, extracellular striated cone; g, glycogen; m, mitochondia; n, nucleus; pb, peribasal body; sg, subacrosomal material; sp, subacrosomal plate; sr,, spiral ridge. Modified from Buckland Nicks and Scheltema (1995).

## Monoplacophora

Spermatozoa of *Laevepilina antartica* presents a low conical acrosomal vesicle, with basal invagination associated to subacrosomal material, The nucleus is a rounded and short with lacunae, the 4 or 5 rounded mitochondrias presses against a dish-shaped depression in the base of the nucleus. The pair of centrioles are arranged in orthogonal and presents a single flagellum (Figure 11) (Healy, 1995)

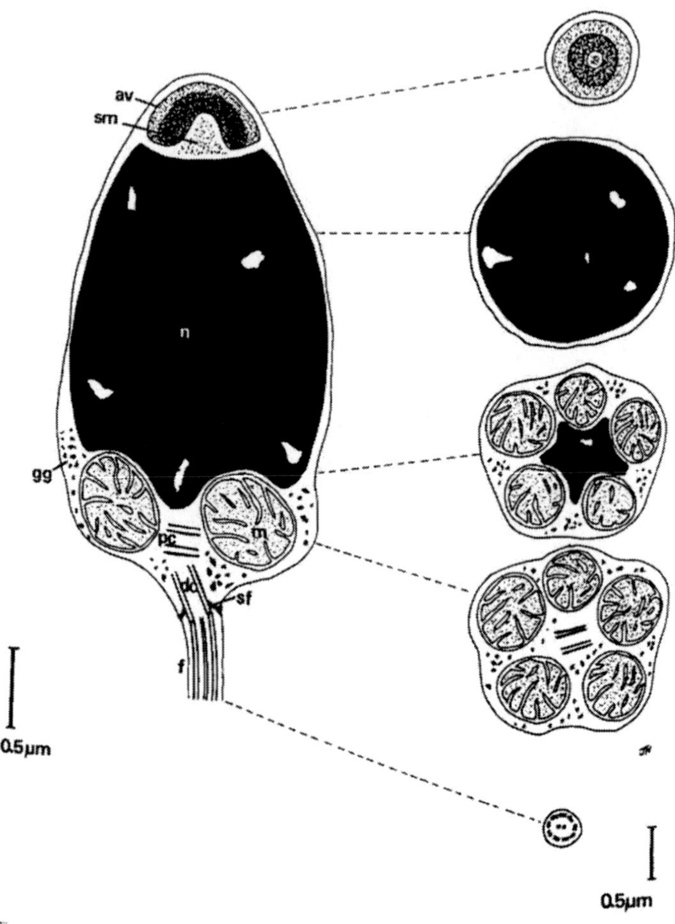


Figure 11. Spermatozoon of the monoplacophoran *Laevipilina antartica*. Abbreviation: av, acrosomal vesicle; dc, distal centriole; f, flagellum; gg, putative glycogen granules; m, mitochondrion; n, nucleus; pc, proximal centriole; sf, satellite fibres; sm, subacrosomal material. (From Healy, 1995).

## Cephalopoda

*Nautilus pompilius* and *Nautilus* present a laterally compressed acrosoma, with a laterally compressed, elongate, rod shaped nucleus, the mitochondrion is situated in each of the two longitudinal grooves of the nucleus.

Octopod cephalopods have screw-shaped spermatozoa, with elongated acrosomes (Franzen, 1967, Longo and Anderson, 1970, Maxwell, 1974).

## Gastropoda

Gastropod exhibits the widest range of sperm morphology to be found in any class of molluscs (Healy, 2000), possible due to the variety of fertilization environments. Comparative studies on the sperm morphology are done in all the groups, and these results are useful as additional characters for phylogenetic analyses. The morphological diversity of spermatozoa in gastropods has been considered as a guide to understand phylogenetic and taxonomic relationships within mollusc.

Within gastropods is possible to find different types of sperms, related to the mode and the capacity of fertilization. The aquasperm type found in most vetigastropods, presents a moderate to large size, conical acrosomal vesicle, a short cylindrical nucleus with a variable number of irregular lacunae, centrioles with an orthogonal arrange, a ring with 4 -5 mitochondrias, nine satellite fibres linking the distal centriole to the plasma membrane and a single flagellum ( Lewis et al., 1980; Sakai et al., 1982; Kohnert and Storch, 1983; Koike, 1985; Healy, 1988a,b; 1990a; Healy and Harasewych, 1992; Hodgson, 1995; Hogdson and Foster, 1992; Hogdson and Chia, 1993).

Fertilization may occur in different scenarios, in the surrounding water (external fertilization), in the female reproductive tract and the mantle cavity (internal fertilization), but it is considered that the last scenario needed a mechanism of sperm transfer.

The typical sperm in caenogastropods is filiform and is described in several studies in the Volutidae family as *Zidona dufresnei* described in Giménez et al., (2008), *Adelomelon ancilla* (Zabala et al., 2009), *Adelomelon beckii* (Arrighetti and Giménez, 2010), *Odontocymbiola magellanica* (Giménez, 2011) and the Olividae family as *Olivancillaria deshayesiana* (Teso and Giménez, 2013).

The euspermatozoa fertile sperm is form by an acrosomal complex, nucleus, midpiece, glycogen piece and end piece with a single incorporated axonemes. The elongated acrosomal complex (Figure 12) continue in a filiform nucleus where the axonemes are inserted in the basal portion (Figure 12, 13).

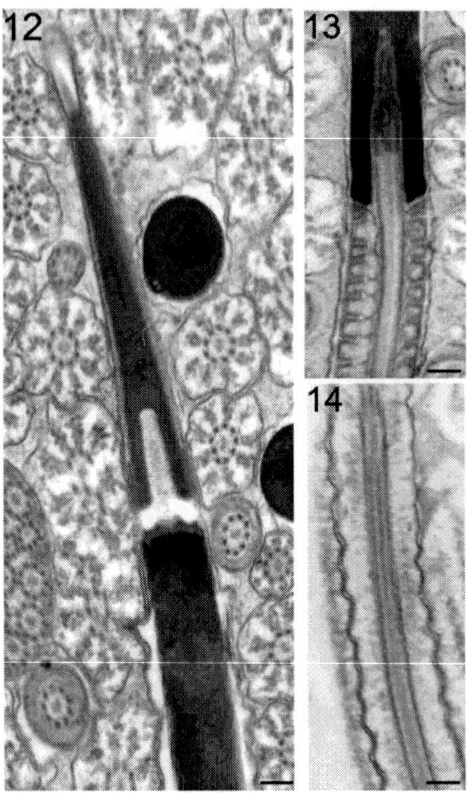


Figures 12-14. Euspermatozoa of *Zidona dufresnei.* 12. Longitudinal section (LS) through acrosomal complex and anterior portion of the nucleus. 13. LS junction of nucleus (showing invagination and centriole/axoneme insertion) and anterior portion of midpiece. Note in LS middle piece elements are spiralling around axoneme. 14. LS glycogen piece. Scale bar 0.5 μm.

## PARASPERM

In addition to the fertile sperm, uniflagellate euspermatozoa (Figure 15), other types of infertile sperms are present in the order Caenogastropoda (Figures 16, 17). Healy and Jamieson (1981) reviewed terminology and proposed the term paraspermatozoa. This infertile sperm have reduced nuclear content (olygopyrene sperm) or no nuclear material (apyrene sperm). These atypical paraspermtaozoa show a large diversity which makes these cells of taxonomic importance.

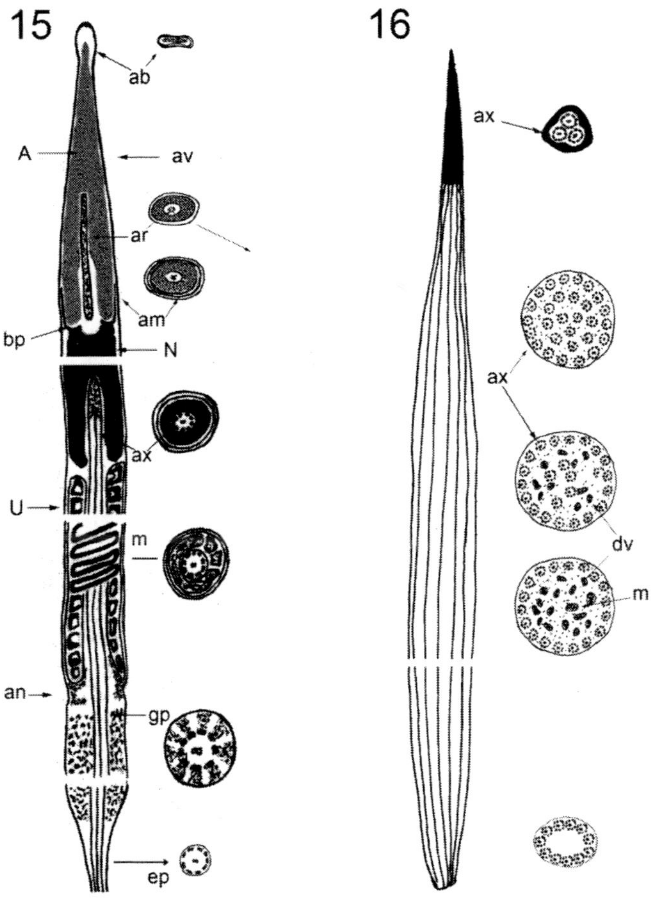


Figures 15-16. Diagram of spermatozoa of the volutid caenogastropod *Adelomelon ancilla*. 15. Internal view of eusperm. TEM. Note the constriction of subacrosomal invagination and the U-shape of electron-dense mitochondrial elements in the midpiece. 16. External and general view of parasperm.Transversal sections observed by TEM. Note the TS of the apical, medial and final region. *A* acrosomal complex; *ab* apical bleb; *ar* axial rod material; *av* acrosomal vesicle; *am* accessory membrane; *an* annulus; *ax* axoneme; *bp* basal plate; *ct* acrosomal vesicle constriction ; *dv* dense vesicles; *ep* endpiece; *g* putative granules; *gp* glycogen piece; *m* mitochondria; *mp* midpiece; *N* nucleus, *U* U-shaped defining edge of mitochondrial element. Not drawn exactly to scale. (From Zabala et al., 2009).

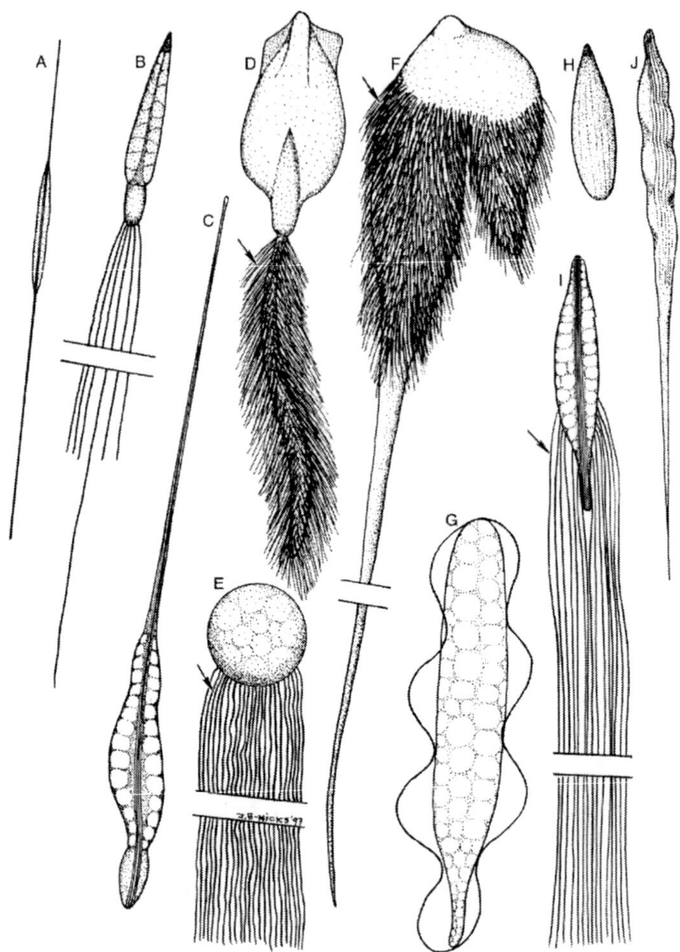


Figure 17. Parasperm major types in gastropods A. *Nerita*; B. *Cerithium*; C. *Serpulorbis*; D. *Janthina*; E. *Littorina*; F. *Epitonium*; G. *Strombus*; H. *Conus*; I. *Fusitriton* lancet sperm (From Buckland-Nicks, 1998).

Giménez et al. (2008) described the morphology of the parasperm in the volutid *Zidona dufresnei* (Figure 18, 19). The paraspermatozoas of *Z. dufresnei* are vermiform with tapered anterior and posterior extremities. In the main body region of the paraspermatozoa is observed: (1) 14–20 (17 ± 3, $n$ = 15) peripherally distributed axonemes lying close to or in contact with the inner surface of the plasma membrane (axonemes approximately equidistant from each other).

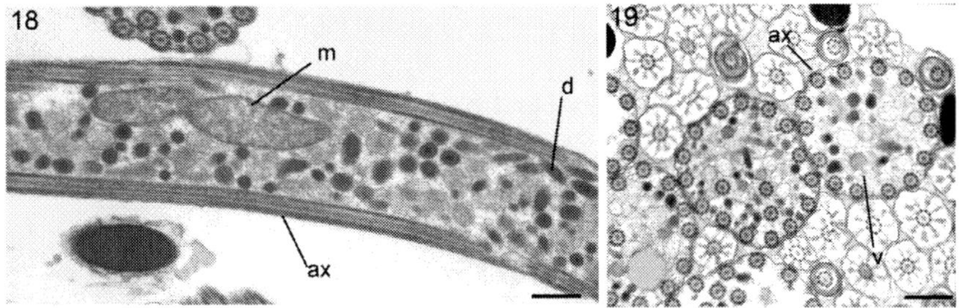


Figures 18-19. Paraspermatozoa of *Zidona dufresnei.* 18. Longitudinal section (LS) main body region of cell showing peripheral axonemes, dense vesicles, less dense vesicles and a large elongate mitochondrion. 19. TS through main body region of paraspermatozoon, showing 15 peripheral axonemes closely adherent to the plasma membrane, dense vesicles and less dense vesicles.. ax axoneme; dv dense vesicles; m mitochondrion; v less-dense. Scale bar 0.5 µm.

In another caenogastropod, from the same family Volutidae, *Adelomelon beckii,* it is possible to observe external features form Scanning electron microscopy (SEM) (Figure 20, 21) of the parasperm described in Arrighetti and Giménez (2010).

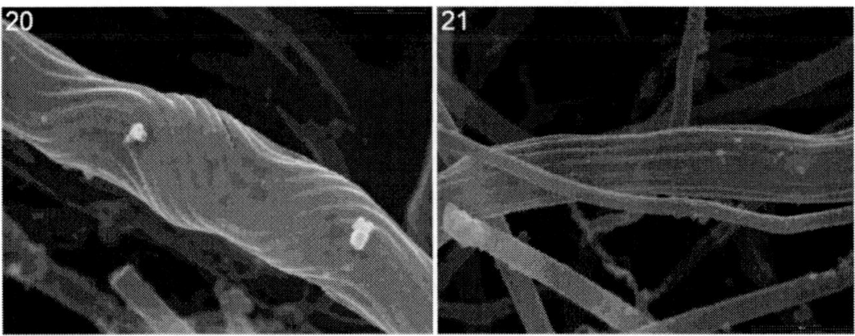


Figures 20-21. Paraspermatozoa of *Adelomelon beckii.* 20. External axonemes in helical way. 21. Axonemes following the straight way. Scale bar 1 µm.

## PARASPERMATOGENESIS

The paraspermatogenic cells are observed in clusters according to their stage of development and are distributed throughout the tubule (Figure 22). It presents larger size than the euspermatogenic cells. Note shorter flagella in posterior section of cell and caryomerites inside the cytoplasm (Figure 22). As seen by optical microscopy, paraspermatid development is characterized by elongation of the cell, concurrent with the appearance of a cytoplasmatic elongation at the apex of the cell and the breakdown of the nucleus into small round fragments (caryomerites) (Figure 22–24). The result of this process is a mature apyrene parasperm of vermiform shape (Figure 25).

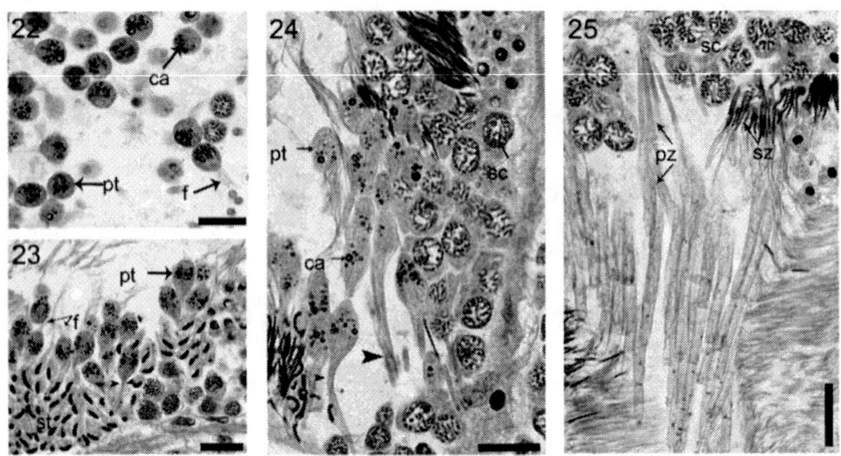


Figures 22-25.Paraspermatogenesis in the volute *Adelomelon ancilla*. 22. Early paraspermatid with short flagella (arrows). 23. Advanced paraspermatid among other euspermatogenic cells. 24. Advanced paraspermatid with an elongated flagellum (arrowhead). 25. Cluster of mature paraspermatozoa. Note their vermiform shape. Abbreviations: bm, basal membrane; ca, caryomerites; f, flagella; pg, paraspermatogonia; pt, paraspermatid; pz, paraspermatozoa; sc, euspermatocytes; sg, euspermatogonia; st, euspermatid; sz, mature euspermatozoa; ser, Sertoli cell. Scale bars: 22–25 = 20 μm. (Modified from Zabala et al., 2012).

During paraspermatogenesis is possible to observe in the early paraspermatid the presence numerous caryomerites as a well-developed Golgi complex and numerous dense granules (Figure 26). In the advanced paraspermatid, the centrioles and centriolar rootlets extends from the core of the axonemes (Figure 27). The apical portion of the parasperm is covered by granular cap of mature paraspermatozoa attached to basal membrane of spermatogenic tubule. A homogenous apical cap covers the axonemes (Figure 30).

The secretory granules present on the parasperm proceed from the activity of the Golgi body. These granules are composed of glycoproteins as shown by PAS-positive reactions (Zabala et al. 2012).

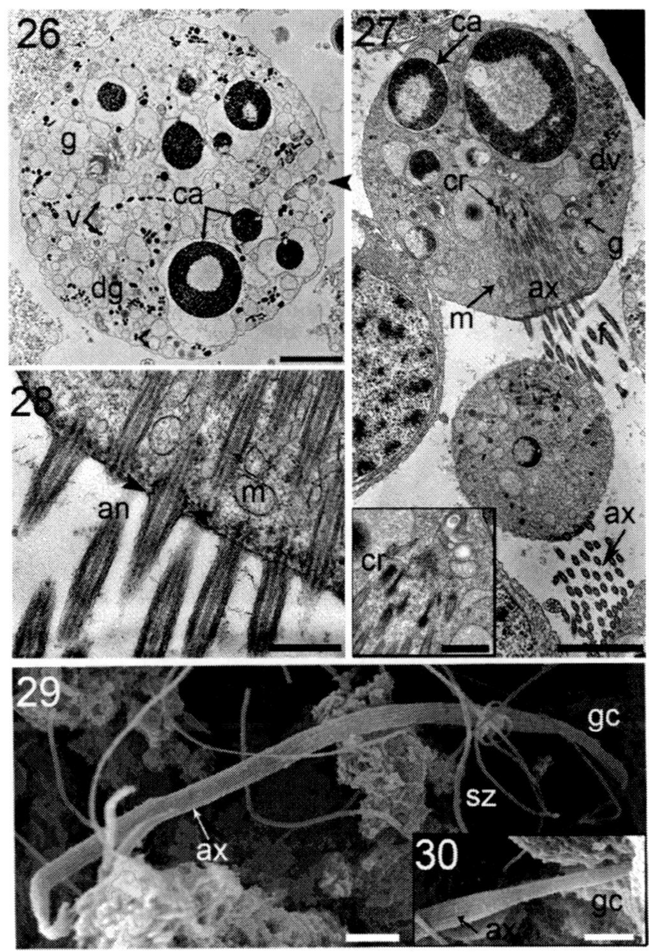


Figures 26-30. Late stages of paraspermatogenesis of *Adelomelon ancilla*. 26, 27, 28. Viewed by TEM. 29, 30. Viewed by SEM. 26. Early paraspermatid. Note vesicle excreted from cell (arrowhead). Note membrane-bound regions enclosing nuclear vesicles, which are pale and similar in density to the many clear vesicles in the cytoplasm. 27. Advanced paraspermatid. Inset: Detail of centriolar rootlets (electron-dense material) of axonemes embedded in an electron-dense material. 28. Detail of the axoneme-annulus. Electron-dense satellite fibres (arrowhead) and mitochondria lie between axonemes. 29. Mature apyrene paraspermatozoa taken from a spermatogenic tubule. Note lancet and vermiform shapes. 30. Detail of apical portion.. A homogenous apical cap covers the axonemes. Abbreviations: an, axoneme-annulus; ax, axoneme; ca, caryomerites; cr, centriolar rootlets; dg, dense granules; dv, dense vesicles; f, flagella; g, Golgi complex; gc, granular cap; m, mitochondrion; sz, spermatozoa; v, nuclear vesicles. Scale bars: 26, 28 = 2 µm; 27, 29, 30 = 5 µm; B inset = 1 µm. (From Zabala et al., 2012).

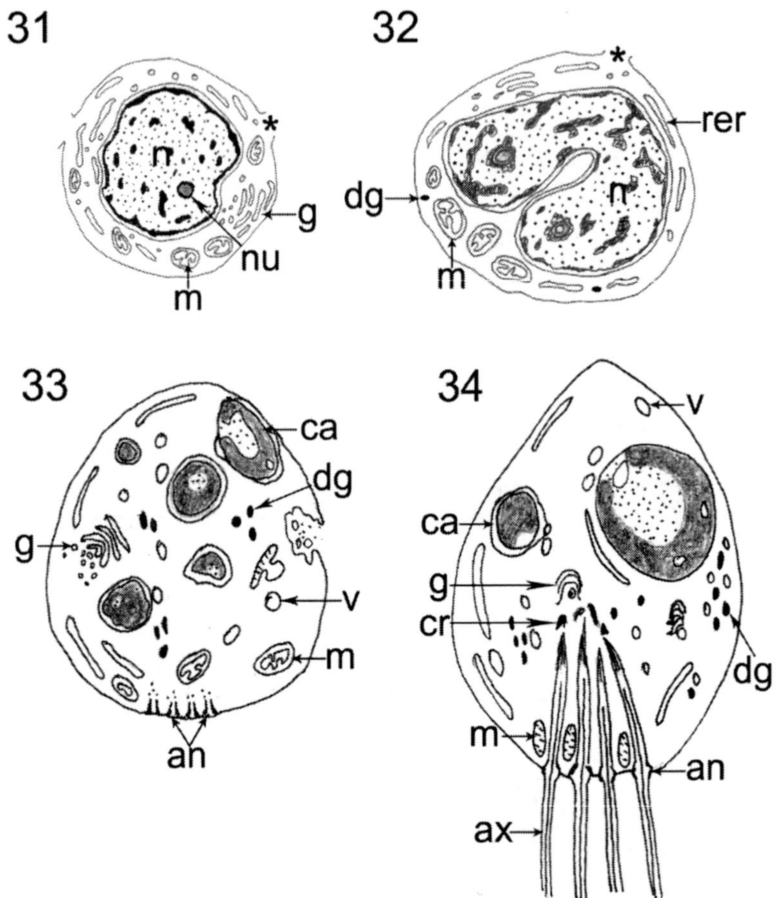


Figures 31-34. Diagram of paraspermatogenesis in *Adelomelon ancilla*. Not exactly to scale. 31. Paraspermatogonium. Note chromatin distribution pattern. 32. Paraspermatocyte. The nucleus invaginates prior to nuclear vesicle formation. 33. Early paraspermatid, with the formation of caryomerites. The basal bodies are attached to the membrane by their annuli. Secretory granules become larger and some, located at the periphery of cytoplasm, discharge their contents by exocytosis. 34. Advanced paraspermatid. Note migration of basal bodies and centriolar rootlets together to the cell apex with consequent development of intracytoplasmic portion of the flagella. Axonemes project through annuli and expand plasma membrane posteriorly to form the flagellar tail brush. Elongated mitochondria are interspersed between adjacent axonemes. Abbreviations: an, axoneme-annulus; asterisk, cytoplasmic bridges; ax, axoneme; ca, caryomerites; cr, centriolar rootlets; dg, dense granules; g, Golgi complex; m, mitochondria; n, nucleus; nu, nucleolus; v, nuclear vesicles. (From Zabala et al., 2012).

These results coincide with the cytochemistry of the apyrene parasperm for a variety of caenogastropods (Buckland-Nicks et al., 1982; Amor and Durfort, 1990). It has been suggested that for caenogastropods in general the granules may attract and bind eusperm to this region of the parasperm (Buckland-Nicks, 1998). Healy and Jamieson (1981) suggested that contents of the vesicles during exocytosis could form a nutrifying fluid for eusperm within the sperm duct or in the genital tract of the receiving individual. However, further studies are necessary to elucidate the exact role of parasperm in the reproductive process.

## REFERENCES

Amor, M. J. & Durfort, M. (1990). Changes in nuclear structure during eupyrene spermatogenesis in Murex brandaris. *Molecular Reproduction and Development,* 25, 348–356.

Arrighetti, F. & Giménez, J. (2010). Ultrastructure of euspermatozoa and paraspermatozoa in the marine gastropod *Adelomelon beckii* (Caenogastropoda, Volutidae). *Helgoland Marine Research*, 64, 143-148.

Buckland-Nicks, J. A. (1998). Prosobranch parasperm: sterile germ cells that promote paternity? *Micron*, 29, 267–280.

Buckland-Nicks, J. A. & Scheltema, A. (1995). Was internal fertilization and innovation of early Bilateria? Evidence from sperm structure of a mollusc. *Proceedings of the Philosophical Society of London,* B, 261, 11-18.

Buckland-Nicks, J. A., Williams, D., Chia, F. S. & Fontaine, A. (1982). The fine structure of the polymorphic spermatozoa of *Fusitriton oregonensis* (Mollusca: Gastropoda), with notes on the cytochemistry of the internal secretions. *Cell and Tissue Research,* 27, 235–255.

Denson, D. R. & Wang, S. Y. (1994). Morphological differences between zebra and quagga mussel spermatozoa. *American Malacological Bulletin,* 11, 79–81.

Denson, D. R. & Wang, S. Y. (1998). Distinguishing the dark false mussel, *Mytilopsis leucophaeata* (Conrad, 1831), from the non-indigenous zebra and quagga mussels, *Dreissena* spp. using spermatozoan external morphology. *Veliger,* 41, 205–211.

Eckelbarger, K. J., Bieler, R. & Mikkelsen, P. M. (1990). Ultrastructure of sperm development and mature sperm morphology in three species of commensal bivalves (Mollusca: Galeommatoidea). *Journal of Morphology, 205*, 63–75.

Franzen, A. (1967). Spermiogenesis and spermatozoa of the Cephalopoda. *Arkiv fur Zoologi*, S 19, 323-334.

Franzen, A. (1983). Ultrastructural studies of spermatozoa in three bivalve species with notes on evolution of elongated sperm nucleus in primitive spermatozoa. *Gamete Res.*, 7, 199–214.

Giménez, J., Healy, J. M, Hermida, G. N, Lo Nostro, F. & Penchaszadeh, P. E. (2008). Ultrastructure and potential taxonomic importance of euspermatozoa and paraspermatozoa in the volutid gastropods *Zidona dufresnei* and *Provocator mirabilis* (Caenogastropoda, Mollusca). *Zoomorphology,* 127, 161–173.

Giménez, J., Arrighetti, F., Teso, V., Hermida, G. N., Zabala, S. & Penchaszadeh, P. E.(2009). Sperm morphology of two marine neogastropods from the southwestern Atlantic Ocean (Caenogastropoda: Volutidae and Olividae). *The Nautilus*, 123, 166–171.

Giménez, J. (2011). Euspermatozoa and paraspermatozoa in the volutid gastropod *Odontocymbiola magellanica* from Patagonia, Argentina. *Acta Zoologica,* 92, 355–362.

Healy, J. M. (1988a). Sperm morphology and its systematic importance in the Gastropoda, in Prosobranch Phylogeny (Ed. W.F. Ponder). *Malacological Review* 4, 251–266.

Healy, J. M. (1988b). Sperm morphology in Serpulorbis and Dendropoma and its relevance to the systematic position of the Vermetidae (Gastropoda). *Journal of Molluscan Studies,* 54, 295–308.

Healy, J. M. (1990). Sperm structure in the scissurellid gastropod *Sinezona* sp. (Prosobranchia, Pleurotomarioidea). *Zoologica Scripta*, 19, 189-193.

Healy, J. M. (1995). Comparative spermatozoal ultrastructure and its taxonomic and phylogenetic significance in the bivalve order Veneroida. In: Jamieson, B.G.M., Ausio, J. and Justine, J.-L. (eds) Advances in Spermatozoal Phylogeny and Taxonomy, Vol. 166. *Mem. Mus. Natl. Hist. Nat.,* Paris, 155–166.

Healy, J. M. (1996). Molluscan sperm ultrastructure: correlation with taxonomic units within the Gastropoda, Cephalopoda and Bivalvia in Origin and evolutionary radiation of the mollusca. (ed. J. Taylor). Oxford University Press, 99-113.

Healy, J. M. (2000). Mollusca: Relict taxa. *Reproductive Biology of Invertebrates, Part B Progress in Male Gamete Ultrastructure and Phylogeny*, pp. 21-79. In. (Series Eds), K.G. R. G. Adiyodi (Series Eds), Jamieson, B. G. M. (Volume Ed.): Vol. IX. pp. Oxford. & IBH Publishing, New Delhi and Calcutta.

Healy, J. M. & Harasewych, M. G. (1992). Spermatozoa and spermatogenesis in *Perotrochus quoyanus* (Fischer and Bernardi) (Gastropoda: Pleurotomariidae) with a description of the male reproductive tract. *The Nautilus,* 106, 1-14.

Healy, J. M., Schaefer, K. & Haszprunar, G. (1995). Spermatozoa and spermatogenesis in a monoplacophoran mollusc, *Laevipilina antartica*: ultrastructure and comparison with other mollusca. *Marine Biology*, 122, 53:65.

Healy, J. M. & Jamieson, B. G. M. (1981). Ultrastructure of spermiogenesis in the gastropod *Heliacus variegatus* (Architectonicidae), with description of a banded periaxonemal helix. *Marine Biology*, 109, 67-77.

Hodgson, A. N. (1995). Spermatozoal morphology of Patellogastropoda and Vetigastropoda (Mollusca: Prosobranchia). In: Jamieson, B.G.M., Ausio, J. and Justine, J.-L. (eds) Advances in Spermatozoal Phylogeny and Taxonomy, Vol. 166. *Mem. Mus. Natl. Hist. Nat.*, Paris, 167–178.

Hogdson, A. N & Chia, F-S. (1993). Spermatozoon structure of some North American prosobranchs from the families Lottidae (Patellogastropoda) and Fissurellidae (Archeogastropoda). *Marine Biology*, 116, 97-101.

Hogdson, A. N. & Foster, G. G. (1992). Structure of the sperm of some South African archaeogastropods (Mollusca) from the superfamilies Haliotoidea, Fissurelloidea and Trochoidea. *Marine Biology*, 113, 89-97

Jamieson, B. G. M. (1987). A biological classification of sperm types, with special reference to Annelids and Molluscs, and an example of spermiocladistics. In: Mohri, H. (ed) New Horizons in Sperm Cell Research. *Gordon and Breach Sci.* Publ., New York, 311–332.

Johnson, M. J., Casse, N. & Le Pennec, M. (1996). Spermatogenesis in the endosymbiont-bearing bivalve *Loripes lucinalis* (Veneroida: Lucinidae). *Molecular Reproduction and Development*, 45, 476–484.

Keys, J. L. & Healy, J .M. (1999). Sperm ultrastructure of the giant clam Tridacna maxima (Tridacnidae: Bivalvia: Mollusca) from the Great Barrier Reef. *Marine Biology*, 135, 41–46.

Keys, J. L. & Healy, J. M. (2000). Relevance of sperm ultrastructure to the classification of giant clams (Mollusca, Cardioidea, Cardiidae, Tridacninae). In: Harper, E.M., Taylor, J.D. and Crame, J.A. (eds) The Evolutionary Biology of the Bivalvia, Vol. 177. *Geological Society*, Special Publications, London, 191–205.

Kohnert, R. & Storch, V. (1983). Ultrastrukturelle Untersuchungen zur Morphologie und Genese der Spermien von Archaeogastropoda |

[Ultrastructural investigations on the morphology and genesis of sperms in Archaeogastropoda]. *Helgoländer Meeresuntersuchungen*, 36, 77-84.

Koike, K. (1985). Comparative ultrastructural studies on the spermatozoa of the Prosobranchia (Mollusca: Gastropoda). Science Report of the Faculty of Education, Gunma University 34, 33–153.

Komaru, A. & Konishi, K. (1996). Ultrastructure of biflagellate spermatozoa in the freshwater clam, *Corbicula leana* (Prime). *Invertebrate Reproduction and Development,* 29, 193–197.

Konishi, K., Kawamura, K., Furuita, H. & Komaru, A. (1998). Spermatogenesis of the freshwater clam *Corbicula fluminea* Muller (Bivalvia: Corbiculidae). *Journal of Shellfish Research*, 17, 185–189.

Lewis, C. A., Leighton, D. L. & Vacquier, V. D. (1980). Morphology of the abalone spermatozoa before and after the acrosoma reaction. *Journal of Ultrastructure Research,* 72, 39-46.

Longo, F. J. & Anderson, E. (1970). Structural and cytochemical features of the sperm of the cephalopds Octopus bimaculatus. *Journal of Ultrastructure Research*, 72: 39-46.

Maxwell, W. L. (1974). Spermiogenesis of *Eledone cirrhosa* Lamarck (Cephalopoda, Octopoda) Proceedings of the Royal society of London, B. 186, 181-190.

Moueza, M. & Frenkiel, L. (1995). Ultrastructural study of the spermatozoon in a tropical lucinid bivalve: *Codakia orbicularis* L. *International Journal of Invertebrate Reproduction and Development*, 27, 205–211.

O'Foighil, D. (1985). Fine structure of *Lasaea subviridis* and *Mysella tumida* sperm (Bivalvia, Galeommatacea). *Zoomorphology*, 105, 125–132.

Ockelmann, K. W. (1964). *Turtonia minuta* (Fabricius), a neotenous veneracean bivalve. *Ophelia*, 1, 121–146.

Ockelmann, K. W. (1965). Redescription, distribution, biology, and dimorphous sperm of *Montacuta tenella* Loven (Mollusca, Leptonacea). *Ophelia,* 2, 211–221.

Perse, J. S. & Woollacott, R. M. (1979). Chiton sperm, no acrosome? *American Zoologist*. 19, 956.

Sakai, Y. T., Shiroya, Y. & Haino-Fukushima, K. (1982). Fine structural changes in the acrosome reaction of the Japanese abalone, *Haliotis discus*. *Development, growth and differentiation*, 241, 531-542.

Sakker, E. R. (1984). Sperm morphology, spermatogenesis and spermiogenesis of three species of chitons (Mollusca, Polyplacophora). *Zoomorphology*, 104, 11-121.

Sousa, M. & Azevedo, C. (1988). Comparative silver staining analysis on spermatozoa of various invertebrate species. *International Journal of Invertebrate Reproduction and Development*, 13, 1–8.

Sousa, M., Corral, L. & Azevedo, C. (1989). Ultrastructural and cytochemical study of spermatogenesis in *Scrobicularia plana* (Mollusca, Bivalvia). *Gamete Research,* 24, 393–401.

Rouse, G. W. & Jamieson, B. G. M. (1987). An ultrastructural study of the spermatozoa of the polychaetes *Eurythoe complanata* (Amphinomidae), *Clymenella sp.* and *Micromaldane sp.* (Maldanidae), with definition of sperm types in relation to reproductive biology. *Journal of Submicroscopy Cytology and Patholo*gy, 19, 573–584.

Teso, V. & Giménez, J. (2013) Sperm morphology of two species of the genus *Olivancillaria* (Gastropoda: Olividae) from the southwestern Atlantic. *Journal of Marine Biological Association of United Kingdom*, 93, 825 - 831

Walker, G. K., Black, M. G. & Edwards, C. A. (1996). Comparative morphology of zebra (*Dreissena polymorpha*) and quagga (*Dreissena bugensis*) mussel sperm: light and electron microscopy. *Canadian Journal of Zoology*, 74, 809–815.

Zabala, S., Hermida G. N. & Giménez., J. (2009). Ultrastructure of euspermatozoa and paraspermatozoa in the volutid snail *Adelomelon ancilla* (Mollusca: Caenogastropoda). *Helgoland Marine Research*, 63, 181-188.

Zabala, S., Hermida, G. N. & Giménez, J. (2012). Ultrastructure of euspermatozoa and paraspermatozoa in the volutid snail *Adelomelon ancilla* (Mollusca; Caenogastropoda). *Journal of Molluscan Studies*, 58, 52-65.

In: Spawning
Editor: Erick R. Baqueiro Cárdenas
ISBN: 978-1-63117-655-5


***Chapter 4***

# REPRODUCTIVE TACTICS OF MARINE SHELLFISH SPECIES FROM NORTH-EASTERN VENEZUELA: ECOPHYSIOLOGICAL IMPLICATIONS

***Luis Freites[1], César Lodeiros[1], Dwight Arrieche[2] and Andrew W. Dale[3]***
[1]Instituto Oceanográfico de Venezuela, Cumaná, Venezuela
[2]Instituto de Investigaciones y Ciencias Aplicadas. Venezuela
[3]GEOMAR, Kiel, Germany

## ABSTRACT

Within the context of the reproductive biology of species, strategies would include the total reproductive model shown by a species (strategies K or R), whereas the reproductive tactics would be variations in the physiological responses that individuals show in response to changes in local environmental conditions superimposed on the typical or basic model. This paper provides a general overview of the studies conducted to date on the reproductive tactics of 14 species of marine bivalves with economic importance distributed in the northeastern region of Venezuela and the importance of environmental variables. These species belong to the Families Arcidae (*Anadara notabilis* and *Arca zebra*), Pectinidae (*Euvola ziczac* and *Nodipecten nodosus*), Ostreidae (*Crassostrea rhizophorae*), Limidae (*Lima scabra*), Mytilidae (*Perna perna*, *P. viridis* and *Modiolus squamosus*), Pteriidae (*Pinctada imbricata*), Pinnidae

(*Atrina seminuda*), Donacidae (*D. denticulatus*), Psammobiidae (*Asaphis deflorata*) and Veneridae (*Tivela mactroides*). All species showed reproductive asynchrony in at least one reproductive annual half cycle, except *E. ziczac* which showed reproductive synchrony in both. These species also showed a combination of opportunistic and conservative tactics in the first and second half of annual cycle. Only *E. ziczac* contrast with this pattern, showing only opportunist tactics that coinciding with the high availability of food between January and April (first pulse of coastal upwelling). In the second reproductive half cycle in June and July, *E. ziczac* undergoes renewed active gametogenesis coincident with a second pulse of coastal upwelling. The information included in this study could be used to formulate better policies to manage fisheries resources, including the periods of fisheries closures and production strategies.

**Keywords:** Bivalve, environmental variables, opportunistic tactics, conservative tactics, synchrony

## INTRODUCTION

Studies of reproduction in bivalve molluscs generate fundamental knowledge on the reproductive biology of these invertebrates and also allow predictions on recruitment to be made. This information is critical with regard to obtaining sufficient seed, establishing closed seasons and determination of minimum catch sizes. More broadly, knowledge gained at this level helps to establish effective resource management and production strategies in aquaculture (Seed, 1976, Arsenault and Himmelman, 1998, Barber and Blake, 1991, Alfaro et al., 2001). In general, a reproductive cycle is defined as the set of events that begin with the provision of the energy required (external or internal), activation of the gonad and gametogenesis, physiological maturity of the already-formed gametes including vitellogenesis, spawning or release of gametes, and finally, the recession of the gonad including the reabsorption of residual gametes and other tissues (Seed, 1976; Barber and Blake, 1991). The triggers for each of these reproductive stages are in turn related to several natural factors that may be both exogenous (e.g., temperature, lunar cycle, food availability and quality, salinity, mechanical stimuli) and endogenous (e.g., genetic, hormonal, energy reserves) (Galtsoff, 1964; Mackie, 1984; Thompson et al., 1996). In this sense, the exogenous factors and genotypic characteristics of the species regulate the timing, duration, intensity and frequency of the reproductive cycle depending on the species and their population.

Thus, individuals of a species that are distributed along latitudinal geographical gradients exhibit a characteristic reproductive pattern with spatial variations driven by the local environmental characteristics (Sastry, 1970, 1979).

Temperature and food availability have been identified as the main environmental variables that regulate the reproductive cycle of marine bivalves (Seed, 1976; Bayne and Newell, 1983; MacDonald and Thompson, 1985a,b; Malachowski, 1988; Jaramillo et al., 1993; Lodeiros and Himmelman, 1994, 2000; Arsenault and Himmelman, 1998; Arrieche et al., 2002). These interactions between endogenous and exogenous factors, in relation to the evolution of each species, generate different reproductive tactics.

Reproductive strategies and tactics bear a hierarchical relationship. Reproductive strategies comprise the total reproductive model shown by a species, such as strategy K (invest significant resources in a few descendants, each of which has a high probability of survival). This strategy can be successful but makes the species vulnerable to the fate of a small number of individuals. On the other hand, strategy R produces many offspring, each of which has a low probability of survival (the species being little dependent on the future of a small number of individuals). Reproductive tactics, on the other hand, would be variations in a population in physiological responses to changes in local environmental conditions superimposed on a typical or basic model (Wootton, 1984). Such changes in reproductive activity are probably the expression of mechanisms of adaptation to environmental changes (Cantillanez et al. 2005; Avendaño et al. 2008).

From the point of view of the available energy source for reproduction, Bayne (1976) classified molluscs as "conservative" species (the energy originates from tissue reserves) and "opportunistic" (the energy comes from the food supply of the environment).

An inverse relationship between carbohydrate levels and the degree of maturity of the bivalve has been observed in conservative species (Navarro et al., 1989). Example of this type of strategy includes the scallop *Chlamys opercularis* (Taylor and Venn, 1979), the mussels *M. edulis* (Gabbott and Bayne, 1973; Dare and Edwars, 1975; Pieters et al., 1979; Zandee et al., 1980), *M. galloprovincialis* (Villalba, 1995) and the cockle *Cerastoderma edule* (Navarro et al., 1989).

In contrast, opportunistic species that undergo gametogenesis in periods of high food availability do not show inverse relationships between reproductive indices and energy reserves and undergo a period of reproductive rest during periods of low food availability.

Examples include the clams *Venus mercenaria* (Ansell et al., 1964), *Macoma balthica* (Beukema and De Bruin, 1977) and the cockle *Glycymeris glycymeris* (Galap et al., 1997). The mussel *M. galloprovincialis* has a continuous opportunistic strategy consisting of successive processes of gonadal restoration and spawning during spring and summer. This does not apparently entail a decrease in energy reserves and likely depends on the abundant availability of food that characterizes such periods (Villalba, 1995).

The Caribbean Sea, being located in the tropics, is generally considered to be biodiverse yet oligotrophic. However, in some coastal tropical regions this is not the case due to organic and nutrient input by rivers, coastal upwelling processes or the influence of trade winds. This results in higher productivities as, for example, in northeast Venezuelan where values may exceed above 231 g C/m2 (Mandelli and Ferraz-Reyes, 1982; Miloslavich and Klein, 2008).

During the upwelling that dominates the first half of the year in northeast Venezuela sea surface temperatures range between 21 and 25 °C.

Phytoplankton biomass is mainly superficial at this time, with the highest density in the first 20 meters and maximum values of 300 cells $ml^{-1}$. Up to 2600 cells $ml^{-1}$ have been measured on occasions, with the highest concentration of chlorophyll *a* of about 8 mg $m^{-3}$ (Varela et al., 2003). These concentrations are adequate to support high productivities of aquatic organisms, thereby not conforming to the idea of low production in tropical systems. These upwelling phenomena in the Caribbean occur mainly in the south-east of the region, including the northeastern continental shelf of Venezuela. It results in high sea fisheries production (increasingly characterized as traditional fishing) that currently supports more than 75 % of the production of about 300 - 400 000 tonnes/year. Molluscs are an important resource here mainly due to the bivalve "pepitona" *Arca zebra*, with approximately 40 000 tonnes / year of which over 90 % originate from Chacopata bank (Figure 1) which has an area of about 70-80 $km_2$ (Lodeiros et al., 2006). In contrast, the upwelling relaxation period is characterized by stratification of the water column, low primary productivity and high sea surface temperatures ranging between 26 and 29 °C (Mandelli and Ferráz, 1982; Ferráz Reyes 1989).

This paper reviews and analyses the reproductive strategies of a total of 14 species of bivalves, representing 10 taxonomic families. These species are mainly found in northeastern Venezuela, particularly in the eastern upwelling ecoregion (Miloslavi), on the north coast of the Araya Peninsula, the Gulf of Cariaco and Mochima Bay in Sucre state and the islands of Cubagua and Margarita in Nueva Esparta state.

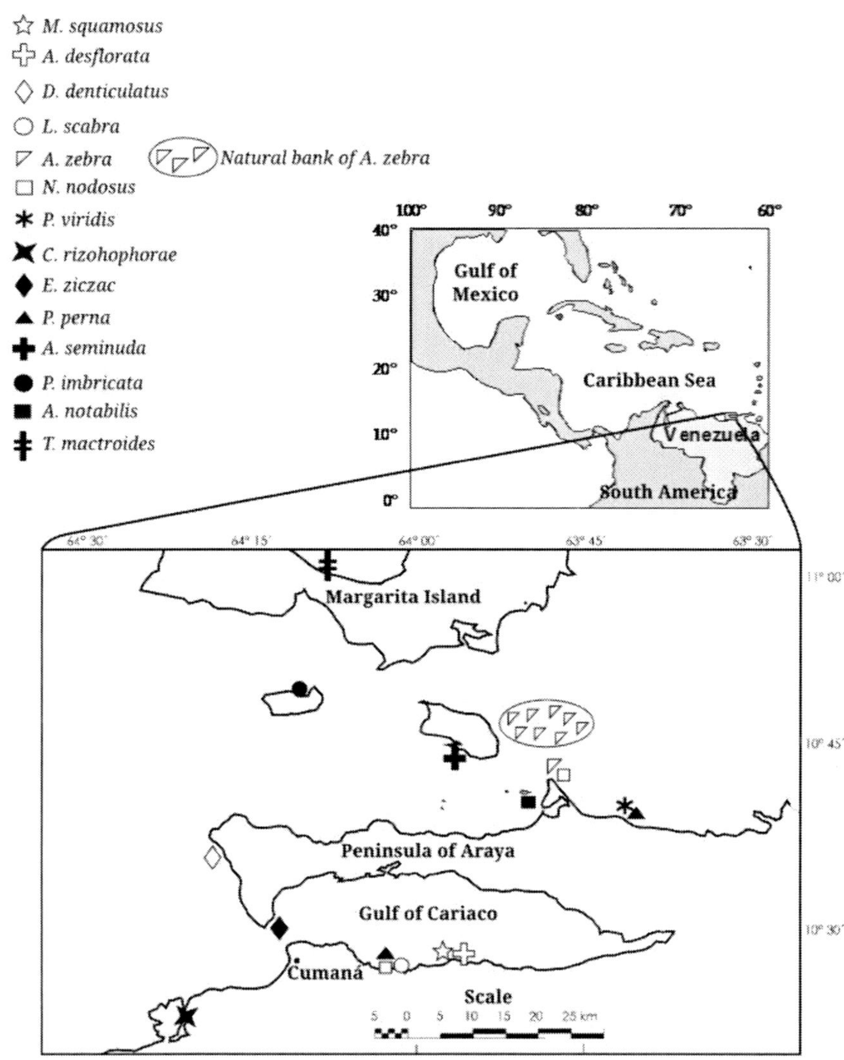


Figure 1. Map of northeastern Venezuela. The symbols indicate the sites where the studies on the species were made.

In one way or another, these areas share the environmental conditions that characterize the ecoregion such as high productivity and environmental variability (Miloslavich and Klein, 2008).

Because this review and analysis is intended to be as comprehensive as possible, each species is listed independently together with all the relevant information.

Thus, although some repetition of physiological characteristics and reproductive tactics is unavoidable, each of the species and their associated reproductive aspects are summarized in Table 1. Reproductive tactics and the environmental conditions associated with different reproductive stages are listed in Table 2. A graphical representation of reproductive cycles and half-cycles of the 14 species of marine bivalve of the coastal upwelling ecoregion (Miloslavich and Klein, 2008) and their relation to temperature and food availability (i.e., chlorophyll *a*) is presented in Figure 3.

## Reproductive Characteristics of the Bivalve Species

### Family Arcidae Lamarck, 1809

*Anadara notabilis* (Röding, 1798). Freites et al. (2010a) reported that *A. notabilis* is a dioecious species (separate males and females) with a female: male sex ratio for the adult population of 1.3:1 (a summary of these features and species analyzed in this study is described in Table 1). With regard to the morphology of the gonad, it is considered to be unobtrusive because it is on the inside of the tissues and fused to the digestive tract. This prevents determination of the sexual status of the gonad using qualitative techniques.

*A. notabilis* shows continuous and asynchronous reproduction with stages of maturity and spawning throughout the year (a graphical representation of the reproductive cycles and half cycles in relation to temperature and food availability for all the species studied is given in Figure 2), but with the highest proportion of mature individuals from March to June and from September to October (Freites et al., 2010a). Moreover, the maximum values in tissue dry mass were observed in March, September and October (Table 2), while the minimum values were observed in June and November coinciding with spawning peaks (January, April, July and November). These results agree with those of Giles (1984).

Freites et al. (2010) also showed that the maximum spawning peaks coincide with a decrease in temperature with a significant inverse relationship, suggesting that temperature is the primary environmental parameter that modulates spawning.

**Table 1. Aspects of reproductive that characterize the bivalve species**

| Specie | Species sex | Sex change | Proportion sexual | Gonad morphology | References |
|---|---|---|---|---|---|
| *A. notabilis* | Dioecious | N.R. | 1.3:1 to ♀ | Visceral mass (Discrete) | Freites et al. (2010a) |
| *A. zebra* | Dioecious | Protandry. Individuals >50 mm are male | 2:1 to ♂ | Visceral mass (Discrete) | Lista et al. (2006) |
| *N. nodosus* | Monoecious (Hermaphrodite) | -. | -. | On intestinal loop (unfused) | Vélez et al. (1987); García et al. (2007); Freites et al. (2003) |
| *E. ziczac* | Monoecious (Hermaphrodite) | -. | -. | On intestinal loop (unfused) | Brea (1986), Vélez et al. (1993) |
| *L. scabra* | Dioecious | Protandry, hermaphroditism in sex change | 1,09:1 | (unfused) | Lodeiros and Himmelman (1999) |
| *P. perna* | Dioecious | N.R. | N.R. | Mantle (Diffuse) | Vélez and Martínez (1967); Arrieche et al. (2002), Narváez et al. (2008) |
| *P. viridis* | Dioecious | N.R. | 1:1.39 to ♂ | Mantle (Diffuse) | Marcano (2004) |
| *M. squamosus* | Dioecious | Protandry. Small males predominate | 1:1 | Visceral mass (Diffuse) | Prieto et al. (1999) |

**Table 1. (Continued)**

| Specie | Species sex | Sex change | Proportion sexual | Gonad morphology | References |
|---|---|---|---|---|---|
| *P. imbricata* | Dioecious | Protandry, From 20 to 48.8 mm dominate ♂, from 48.8 to 99 mm dominate ♀ | N.R. | Visceral mass (Diffuse) | Marcano (1994) |
| *C. rhizophorae* | Dioecious | Protandry, hermaphroditism | | Visceral mass (Diffuse) | Vélez, 1977, 1991, *Nascimento (1978), |
| *A. seminuda* | Dioecious | Hermaphroditism | 1:1 | Fused to hepatopancreas (Diffuse) | Soria et al. (2002); Freites et al. (2010b) |
| *A. deflorata* | Dioecious | Possible protandry | 1:1 | Visceral mass (Diffuse) | Berg and Alatalo, 1985; Prieto, 2008 |
| *D. denticulatus* | Dioecious | - | 1:1 | Visceral mass (Diffuse) | Vélez, 1985 |
| *T. mactroides* | Dioecious | Protandry, hermaphroditism | 1:1 | Visceral mass (Diffuse) | Prieto, 1980; 1983 |

N.R.: not reported. [a]Discrete: Gonad not observable upon separating the valves or other tissues; Diffuse: Gonad observable upon separating the valves but fused to other soft tissues; Unfused: Gonad observable upon separating the valves and easily removable from the other tissues.

**Table 2. Reproductive tactics, environmental conditions and different reproductive stages of bivalves**

| Species | Reproduct. stage | Period | Environment | Reproduct. tactic | References |
|---|---|---|---|---|---|
| *A. notabilis* | Gametog. and maturity almost year-round | Highest % of mature individuals in March-June and September-October | ▲/▼F | Op/Con-Continuous Asyn | Freites et al. (2011a)* |
| | Spawning | At least 5 periods (January, April, July, September, November) | ▲/▼T ▲/▼F | Asyn | |
| *A. zebra* | Gametog. and maturity almost year-round | Highest % of mature individuals in January, July-September and November | ▲/▼F | Op/Con-Continuous Asyn | Lista et al. (2008) |
| | Spawning | At least 2 periods (February-May, October-December) | ▲/▼T ▲/▼F | Asyn | |
| *P. viridis* | Gametog. and maturity | Highest % of mature individuals in November-December and until July | ▲/▼F | Op/Con- Asyn | Marcano (2004) |
| | Spawning | Spawning between December-January and until July | ▲/▼T ▲/▼F | Asyn | |
| *P. imbricata* | Gametog. and maturity | Highest % of mature individuals in March-April and July-September | ▲/▼F | Op/Con- Asyn | Marcano (1984) |
| | Spawning | At least 3 periods: May-July, September-October, December-February | ▲/▼T ▲/▼F | Asyn | |

**Table 2. (Continued)**

| Species | Reproduct. stage | Period | Environment | Reproduct. tactic | References |
|---|---|---|---|---|---|
| *P. perna 1st reproduc.half cycle* | Gametog. and maturity and partial spawning | March-May | ▼T+▲F | Op Asyn | Vélez and Martínez (1967); Arrieche et al. (2002); Narváez et al. (2008)*; |
| | Spawning | June-July | ▲T+▲F | Asinc. | |
| *2nd reproduc. half cycle* | Gametogenesis and maturity | August-December | ▲T+▼F | Con-Asyn | |
| | Spawning principal | January-February | ▼T+▲F | Mainly Syn | |
| *C. rhizophorae* | Gametog. and maturity almost year-round | Highest % of individuals in gametogenesis from January-May. Maturity May-November | ▲/▼F | Mainly Con and Oport Asyn | Vélez (1977, 1991) Marcano (1984) |
| | Spawning | Partial spawning all year with peaks in July-November. At least 5 periods. | ▲T/▼T ▲/▼F | Asyn | |
| *A. deflorata 1st reproduc. half cycle* | Gametog. Maturity | February-May Julio | ▼T+▲F | Op Asyn | Prieto et al. (2008); |
| | Spawning principal | August-September | ▲T+▲F +▼Sal. | Asyn | |
| *2nd reproductive half cycle* | Gametogenesis and maturity | November-December | ▲T+▼A | Op-Asyn | |
| | Spawning | January | ▼T+▲A | | |

| Species | Reproduct. stage | Period | Environment | Reproduct. tactic | References |
|---|---|---|---|---|---|
| *A. seminuda 1st reproduc. half cycle* | Gametog. and maturity<br>Spawning | January-April<br>May-June | ▼T+▲F<br>▲T+▲A | Op- Asyn<br>Asyn | Freites et al. (2001b)* |
| *2nd reproduc. half cycle* | Gametog. and maturity<br>Spawning | July-August and October-November<br>(September, December, and January) | ▲T+▼A<br>▲T+▼A<br>▼T+▲A (January) | Con- Asyn<br>Asyn | |
| *L. scabra 1st reproduc. half cycle* | Gametog.<br>Maturity<br>Spawning | January-April<br>May<br>June-July | ▼T+▲F<br>▲T+▼F<br>▼T+▲F | Op-Asyn<br>Asyn<br>Syn | Lodeiros and Himmelman (1999) |
| *2nd reproduc. half cycle* | Gametog. and maturity<br>Spawning | August-September<br>October-December | ▲T+▼A<br>▼T+▲F | Con-Asyn<br>Syn | |
| *M. squamosus 1st reproduc. half cycle* | Gametog.<br>Maturity<br>Spawning | December-April<br>May<br>June-July | ▼T+▲A<br>▲T<br>▼T+▲A | Op-Asyn<br>Asyn<br>Asyn | Prieto et al. (1999) |
| *2nd reproduc. half cycle* | Gametog. and maturity<br>Spawning | August-September<br>October | ▲T+▼A<br>▼T+▲A | Con-Asincr<br>Asyn | |
| *N. nodosus 1st reproduc. half cycle* | Gametog., maturity and partial spawning<br>Madurez and Spawning | January-April<br>May | ▼T+▲F.<br>▲T+▲F | Op-Asinc.<br>Asyn | Velez et al. (1987); García et al. (2007); Freites et al. (2003)* |

**Table 2. (Continued)**

| Species | Reproduct. stage | Period | Environment | Reproduct. tactic | References |
|---|---|---|---|---|---|
| *2nd reproductive half cycle* | Gametog. and maturity<br>Spawning | August-October<br>November-December | ▲T+▼A<br>▼T+▲A | Con- Syn<br>Syn | |
| *E. ziczac 1st reproduc. half cycle* | Gametog.<br>Maturity<br>Spawning | December-April<br>May<br>May-June | ▼T+▲A<br>▲T<br>▼T | Op-Syn<br>Syn<br>Syn | Brea (1986), Vélez et al. (1993) |
| *2nd reproductive half cycle* | Gametog.<br>Maturity<br>Spawning | June-July<br>August<br>September | ▼T+▲A<br>▲T<br>▲T | Op-Syn<br>Syn<br>Syn | |
| *D. denticulatus 1st reproducc. half cycle* | Gametog., maturity and spawning | January-July | ▼T+▲A | Op-Asyn | Vélez (1985) Prieto et al. (1999) |
| *2nd reproduc. half cycle* | Gametog. and maturity<br>Spawning | September-October<br>October-November | ▲T+▼A<br>▲T+▼A | Con Asyn<br>Asyn | |
| *T mactroides* | Gametog.<br>Maturity<br>Spawning | March-June<br>July-November<br>December | ▼T+▲A<br>▲T +▼A<br>▼T+▲A | Op-Asyn<br>Asyn<br>Asyn | Prieto et al. (1983) |

Op = opportunistic; Con = Conservative; Syn = Synchronous; Asyn = Asynchronous; ▲T = Temperature increase; ▼T = Temperature decrease; ▲F = Food increase; ▼F = Food decrease; ▼Sal. = Salinity decrease.

*Arca zebra* (Swainson, 1833). Lista et al. (2008) showed that *A. Zebra* is a dioecious species. However, one can observe a ratio of almost 100% of males in individuals <50 mm in anteroposterior length, which conditions protandric development. *A. zebra* also presents a sex ratio for the population at large (> 70 mm) of varying proportions with periods of alternating high percentage of males and females, suggesting a possibility of alternating hermaphroditism. The gonad is discrete (fused with the digestive tissue), making it problematic to determine the sexual status of the gonad using qualitative techniques.

Most research shows that gametogenesis appears to be continuous but peaks in July, October and March (Mora, 1985, Alvarez, 1992, Saint-Aubyn et al., 1992). Lista et al. (2008) report that periods with the highest proportion of spawned individuals occurs between February to April and October to January (Table 2) and that maturity stages occur either in January, July-September and November (Lista et al., 2006, 2011). These periods are characterized by either low or high food availability of phytoplankton origin.

## Familia Pectinidae Rafinesque, 1815

*Nodipecten nodosus* (Linneo, 1758). Hermaphrodite species, with an easily discernible gonad such that with the naked eye and some experience one can make a classification of the different sexual stages based on the colour and size of the gonads without dissecting the body. The gonad is almost independent of the rest of the soft tissues since it is located an intestinal loop with the female part a reddish-orange colour and male part being white to creamy white.

The following description is based on studies by Vélez et al. (1987), Freites et al. (2003) and Garcia et al. (2010). Although a greater degree of reproductive asynchrony in the population has been observed in the first half of the year with partial spawning, this species generally has active gametogenesis between January and April (Table 2). Maturation of existing gametes and spawning begins in late April and May when temperatures regularly exceed 25 °C.

Thereafter, the species undergoes a period of sexual rest in parallel with the accumulation of energy reserves. This is supported by the high availability of phytoplankton-based food that is produced in the second pulse of upwelling which occurs during June and July in the southeastern Caribbean (Muller-Karger and Varela, 1988). From August to October a new process of active gametogenesis begins.

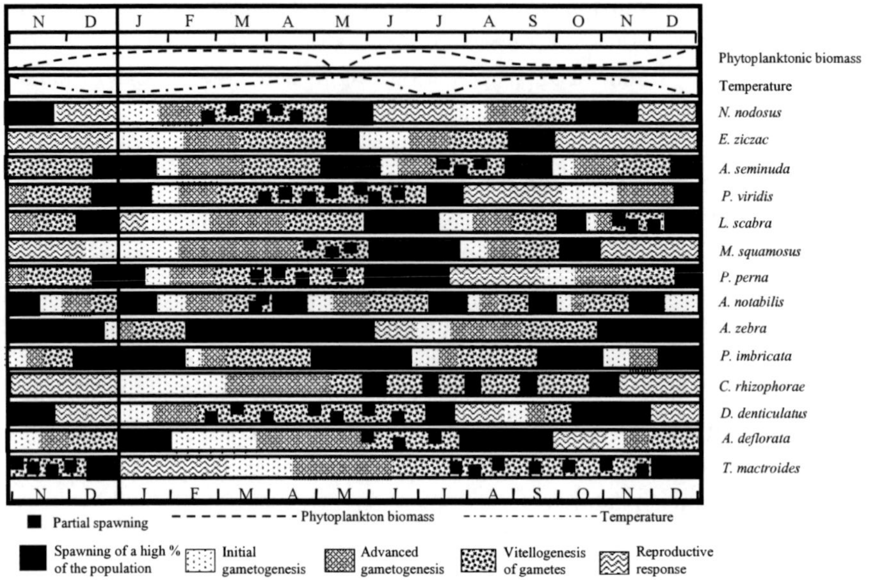


Figure 2. Variations in the annual cycles of half cycles of 14 species of marine bivalves of the coastal upwelling ecoregion and the relation to water temperature and food availability of phytoplanktonic origin (i.e., chlorophyll *a*).

The process of maturation of previously-formed gametes occurs when coastal upwelling is re-initiated between November and December (depending on the interannual variability) when temperature decreases to below 25 °C. Once gamete maturation is complete, a second spawning period follows. The second reproductive half cycle is considered, from the standpoint of population, more synchronous and therefore more intense since a very high percentage of the population participates. Furthermore, this half cycle is considered to be conservative since it depends on the energy reserves accumulated in muscle (Garcia et al., 2007) and in the digestive gland (Freites et al., 2010). *Euvola ziczac* (Linneo, 1758). Hermaphrodite species with a readily discernible gonad. As in *N. nodosus*, it can be observed with the naked eye for the classification of different sexual stages. The gonad is almost independent of the rest of the soft tissues since it is located on a loop of intestine, with the female part being light orange to reddish orange while the male part is white to creamy white. This is one of the species of the region where a greater degree of reproductive synchrony has been observed since in the two spawning periods a high percentage of individuals in the populations studied was involved.

The following description is based on studies by Brea (1986) and Velez et al. (1993).

Generally, one can say that this species has active gametogenesis between January and April, while the process of maturation of the formed gametes begins in late April and May when temperatures regularly exceed 25 °C (Table 2). After this process of gamete maturation a spawning period occurs involving a high percentage of the population.

Subsequently, and unlike *N. nodosus*, it goes directly to a process of active gametogenesis (virtually no sexual rest), taking advantage of the relatively high availability of phytoplankton-based food produced in the second pulse of coastal upwelling during June and July (Muller -Karger and Varela, 1988).

The process of maturation of formed gametes occurs when temperatures rise above 25 °C, usually from August with high frequency spawning in September. Sexual rest follows until November -December. The association of reproductive half cycles with the primary and secondary coastal upwelling pulses suggests an opportunistic reproductive strategy.

Thus, the species profits from the food supply offered by the environment to channel some of the energy directly into gametogenesis.

## Familia Limidae Rafinesque, 1815

*Lima scabra* (Born, 1778). Dioecious, protandrous species exhibiting hermaphroditism with a ratio between males and females of 1.1:1 (Lodeiros and Himmelman, 1999). It has an easily discernible gonad that is, as in the scallop, observed with the naked eye. This organ is almost independent of the rest of the soft tissues since it is located on a loop of intestine with the female part being purplish gray.

Although in the first half of the year a greater degree of reproductive asynchrony in the population of *L. scabra* and partial spawning events have been observed, in general it can be stated that this species has an active gametogenesis period between January and April while the process of maturation of formed gametes begins in late April and May when temperatures exceed 25 °C (Table 2).

After this process of gamete maturation, a spawning period occurs involving a high percentage of the population and is induced by the drop in temperature during the second pulse of coastal upwelling (Figure 2).

Subsequently, the species passes a reproductive resting stage in parallel with the accumulation of energy reserves during the months of June and July.

From August to October a new active gametogenesis process seems to occur which takes place under ambient conditions characterized by low availability of phytoplankton-derived food and temperatures close to 28 ºC (Mandelli and Ferraz-Reyes, 1982). The process of maturation of formed gametes begins when coastal upwelling restarts, indicated by the drop in temperature to below 25 ºC that usually occurs between November and December depending on the interannual variability.

Once the process of gamete maturation is completed, a second spawning period begins with a greater degree of reproductive synchrony that is always associated with a temperature decrease.

## Familia Mytilidae Rafinesque, 1815

*Perna perna* (Linneo, 1758). This species is dioecious, with male-to-female ratio of 1:1.39 (Velez and Martinez, 1967). The gonad is diffused into the mantle, yet it is possible to observe macroscopically by sacrificing individuals. Alternatively, the sex and sexual maturity can be determined without sacrificing the individual by puncturing the mantle.

Although January and February is the main spawning period, this process begins in the previous year and, therefore, is considered in this study to be part of the second annual half cycle.

Generally, the active gametogenesis period is between March and May (Arrieche et al., 2002, Acosta et al., 2010), accompanied by a partial spawning of the population. However, in June-July a greater percentage of the population participates (Acosta et al., 2010).

After this first half cycle, the second half cycle begins immediately with active gametogenesis from August to December, reaching the highest percentage of reproductive maturity in December. The reproductive process culminates in the first two months of the following year with another spawning period characterized by the participation of a large percentage of the population. However, Arrieche et al. (2002) observed the minimum gonosomatic index values for this species in May.

In both half cycles, an asynchronous-type reproductive activity is observed. However, the second half cycle that develops in much of the period of low food availability and high temperatures (conservative tactic) is relatively less asynchronous since it culminates with a higher percentage (> 60%) of individuals in the population in the mature and spawning stages (Acosta et al., 2010).

*Perna viridis* (Linneo, 1758). This species is dioecious, with male-to-female ratio of 1:1.4 (Marcano, 2004). As with *P. perna*, its gonad is diffused into the mantle.

The following description is based on work by Marcano (2004) and Acosta et al. (2010) and corresponds to the first annual half-cycle of this species. *P. viridis* develops a process of active gametogenesis in February and March and a spawning peak in April (45% of the population). After this first half cycle, the second immediately begins with active gametogenesis between August and January, reaching the highest percentage of reproductive maturity in January, culminating with spawning in January (Acosta et al., 2010).

However, Marcano (2004) reported spawning in part of the population in December of the same year. Furthermore, asynchronous reproductive activity was observed in the first half cycle while the second half cycle that largely develops during the period of low food availability and high temperatures is relatively more synchronous since it culminates in a greater percentage of individuals at 60% maturity stage (Acosta et al., 2010).

*Modiolus squamosus* (Beauperthuy, 1967). This mussel is a dioecious species, exhibiting protandry. There is a predominance of males in smaller sizes with a 1:1 sex ratio in larger sizes (Prieto et al., 1999). The gonad is diffuse and is closely linked to the mantle.

This species presents active gametogenesis between December and April with some partial spawning of the population. However, in June-July the participation of a greater percentage of the population is observed, which could be characterized as a major spawning peak (Prieto et al., 1999). After this first half cycle, the second immediately begins with active gametogenesis between August and September, reaching the highest percentage of reproductive maturity in September. The process culminates with another spawning period in October. In the first half cycle asynchronous reproductive activity is observed, while the second is relatively more synchronous since it culminates in a greater percentage of individuals in sexual maturity stage. Furthermore, the sum of individuals in partial and total spawning stages can exceed 50% (Prieto et al., 1999).

## Familia Pteriidae Gray, 1847 (1820)

*Pinctada imbricada* (Röding, 1798). This bivalve is dioecious, with initially protandric individuals since the largest proportion of individuals with sizes <50 mm are males. The gonad is closely linked to the visceral mass and is therefore diffuse (Marcano, 1984).

The species has a nearly continuous reproduction during the year (Marcano, 1984, Ruffini, 1984), with several periods of active gametogenesis, maturation and spawning (half cycles) under conditions of both food abundance and scarcity and low and high temperatures.

The highest percentages of mature individuals are observed in the periods from March to April and from June to September, while at least three major spawning periods have been observed from May to June, September to October and December to February.

It is therefore a species with almost continuous reproduction, with an opportunistic or conservative reproductive tactic depending on the availa-bility of food. This implies adaptive flexibility with asynchronous sexual behaviour.

## Familia Ostreidae Rafinesque, 1815

*Crassostrea rhizophorae* (Guilding, 1828). This bivalve is dioecious with protandry in sizes between 20 and 48 mm. In the largest sizes female individuals predominate, with a diffuse gonad since it is closely linked the visceral mass (Velez, 1977, 1991).

It exhibits gametogenesis and mature individuals all year, with the highest percentage of individuals from January to April-May and mature individuals from May to November.

On the other hand, it exhibits spawning during much of the year with peaks from July to November and sexual rest in December (Velez, 1977, 1991). The reproductive tactic is conservative and opportunistic, depending on food availability, and is also asynchronous and almost continuous for much of the year.

## Familia Pinnidae Leach, 1819

*Atrina seminuda* (Lamarck, 1819). This bivalve is dioecious with hermaphroditism in 1% of the population (Soria et al., 2002). The gonad is diffuse since it is closely linked to the hepatopancreas.

This species has a nearly continuous reproduction during the year (Freites et al., 2010b), with gametogenesis and maturity under conditions of food abundance and scarcity of phytoplankton and low or high temperatures.

The highest percentages of mature individuals are observed in the periods from March to May and from June to September (Table 2), while at least three major spawning periods have been observed in the periods from May-June, September-October and December-January.

It therefore has, like *Pinctada imbricata*, high adaptive flexibility with almost continuous reproduction and opportunistic or conservative tactics, depending on the availability of food and asynchronous sexual behaviour.

## Familia Donacidae Fleming, 1828

*Donax denticulatus* (Linneo, 1758). Velez (1985) show that sex ratio was 1:1 in individuals with sizes >9 mm. Sexual differentiation was observed in 50% of juveniles of approximately 9 mm long.

Gametogenesis begins asynchronously (Velez, 1985).

Adults are reproduced continuously and throughout the year, but with greater activity between July and December when the water temperature reaches 28°C.

## Familia Psammobiidae Fleming, 1828

*Asaphis deflorata* (Linneo, 1758). A dioecious species with a sex ratio of 1:1 for clams 25-70 mm shell length. All individuals above 25 mm reached reproductive maturity.

The gonads in mature clams penetrate the foot and the viscera that surround the digestive tubules (diffuse condition). Females are fertile, producing an estimated $5x10^4$ to $5x10^5$ eggs during the breeding season that are released in a single mass spawning (Berg and Alatalo, 1985).

This species breeds throughout the year. The first annual half cycle begins in February-March with an active gametogenesis.

Gonadal maturity peaks in July with spawning in August-September when temperature increases and salinity decreases. Following spawning, the gonads are empty in most adults during the period from September to October.

A new gametogenic cycle begins in November with a peak in gamete maturation and spawning in January (Prieto et al., 2008; Prieto, pers. Comm.).

### Familia Veneridae Rafinesque, 1815

*Tivela mactroides* (Börn, 1778). Dioecious species but isolated cases of hermaphroditism may be found. The total sex ratio for the entire population is 1:1. However, males dominate (protandry) in individuals < 25 mm long; between 25 and 30 mm there is a balance of the sexual ratio; and from 30.1 to 45 mm females dominate (Prieto, 1980). There is asynchrony in sexual maturation. The bivalve reproduces all year but there is a predominance of mature oocytes from July to December when spawning probably occurs in pulses. Juvenile recruitment begins at this time. The maturation of gametes decrease from January to March. Apparently, temperature variation, nutrient cycling and salinity have an important role in reproduction, although mature individuals are found year-round (Prieto, 1980, 1983).

## Discussion

The reproductive system of many families of bivalves is closely related to the digestive system (Mackie, 1984) although there are different degrees of this "corporal approximation". In our analysis (Table 1), *Euvola ziczac* and *Nodipecten nodosus* (Family Pectinidae) and *Lima scabra* (Family Limidae) show one of the most few discrete gonads. Like all pectinids, their gonad is tongue-like and is quite separate from other soft tissues (Beninger and Le Pennec, 2006). An intestinal loop that passes through the gonad tissue.

In the Ostreidae family (which includes *Crassostrea rhizophorae*) this tissue is less separated and diffusely overlays the digestive system (Andrews, 1979) separated by a basement membrane and connective tissue. The Pteriidae family (which includes *Pinctada imbricata*) and most other families of bivalves, present discrete gonads. The gonadal tissue originates from the mesoderm, gradually develops and disseminates around the digestive gland and then mixes with other organs of the visceral mass (Saucedo and Southgate, 2008). The gonadal tissue in *Tivela mactoroides* (Veneridae), *Anadara notabilis* and *Arca zebra* (Arcidae), *Asaphis deflorata* (Psammobiidae) and *Donax denticulatus* (Donacidae) lines the alimentary canal, entering between the lobes of the hepatopancreas. It can often extend into the inner foot, as has been anatomically described for the venerid *Mercenaria mercenaria* (Eble, 2001). Similarly, the gonad in *Atrina seminuda* (Pinnidae) is closely linked to the hepatopancreas which it almost completely surrounds (Freites et al. 2010b).

In this case, however, it is more conspicuous than in the venerids, and can be easily observed by separating the two shells (Soria et al., 2002). In the family Mytilidae (which includes mussels *Perna perna*, *Perna viridis* and *Modiolus squamosus*), the gonadal alveoli diffuse into the mantle tissue and to the edge of the visceral mass (White, 1937).

From the 14 commercially important species included in this study, 10 showed a reproduction almost continuously throughout the year (*P. viridis*, *A. notabilis*, *A. zebra*, *N. nodosus*, *C. rhizophorae*, *P. imbricata*, *A. seminuda, T. mactroides, A. desflorata and D. denticulatus*), whereas the other 4 showed two reproductive half cycles (*E. ziczac*, *L. scabra*, *P. perna* and *M. squamosus*). The behaviour of the 10 first species corresponds to the classic pattern of continuous reproduction of tropical bivalves molluscs (Giese and Pearse, 1974, Vélez, 1977, Sastry, 1979). These species, along with *P. perna* and *M. squamosus*, presented asynchronous reproduction because individuals from different reproductive stages were always reported during a single sampling. As with most species analyzed in this study, several species of bivalves from tropical regions show a continuous and asynchronous reproduction whereby individuals in gametogenesis, mature and spawning stages have been observed in every month of the year. Examples in addition to the species studied here include *Anadara grandis* (Cruz-Soto, 1987), *Polymesoda radiata* (Ruiz et al., 1998) and *Atrina maura* (Angel-Perez et al., 2007). The species studied here are all found in the coastal upwelling ecoregión creating intrinsic conditions that makes it different from a typical tropical region. The region is characterized by two periods with differing levels of primary productivity and water temperature due to wind-driven upwelling and stratification. This generates significant environmental variability that could influence the reproduction of invertebrates (Lodeiros and Himmelman 1994). Nonetheless, the evolutionarily-conditioned reproductive strategies of the described species for the region generally tend to be associated with the relatively small seasonal changes that characterise tropical regions (Giese and Pearse, 1974), such as the tropics (Joseph and Madhyastha, 1984; Lasiak, 1986; Shafee, 1989; Fournier, 1992; Frenkiel, 1997) and deepwater (Le Pennec and Beninger, 1997). This contrasts with bivalves from temperate waters, where the reproductive process manifests itself in a period restricted to one or two seasons with a high degree of synchrony (Mackie, 1982). The timing in the reproductive stages of species of temperate and boreal regions is the result of adaptation to changes engendered by distinct seasonal variability in physiologically important environmental factors such as temperature and food availability.

Examples of this reproductive behaviour include *Dreissena polymorpha* (Haag and Garton, 1992), *Nuculoma tenuis* (Harvey and Gage, 1995) and *Eurhomalea exalbida* (Morriconi et al., 2002).

Of all the species studied here, *E. ziczac and L. scabra* showed reproductive synchrony which could be related to its subtropical distribution (Lodeiros et al. 1999) and, therefore, are evolutionarily adapted to the type of synchronously reproduction typical of habitats with greater environmental variability.

The scallop *Euvola ziczac* showed two periods of active gametogenesis during the upwelling period characterized by high availability of food and therefore considered as purely opportunistic.

In general, the other species combine the two reproductive tactics (opportunistic and conservative), in particular *A. notabilis, A. zebra, N. nodosus, L. scabra, P. perna, P. viridis, M. squamosus* and *D. denticulatus*.

During upwelling (January to July), characterized by high availability of phytoplankton-based food and temperatures below 25 °C (Mandelli and Ferraz-Reyes, 1982), these species showed a clear opportunistic reproductive tactic.

In the second half of the year, characterized by a relatively low food availability in quantitative and qualitative terms (Pirela et al., 2008) and temperatures above 25 °C (Mandelli and Ferraz-Reyes, 1982), they showed more conservative reproductive tactics (sensu Bayne, 1976).

At present, it has been demonstrated that the conservative and opportunistic categories are not exclusive since many species adopt a combination of both to ensure reproductive function (Racotta et al., 1998, 2003, Kang et al., 2000; Luna Gonzalez et al., 2000; Arellano-Martinez et al., 2004; Vite-Garcia, 2005; Vite-Garcia and Saucedo, 2008, Gomez-Robles and Saucedo, 2009; Freites et al., 2010a, b).

This tactic, that includes both reproductive behaviours in an annual period, occurs mainly as a response to changes in water temperature and food availability in a specific location, which in turn often depends on the microclimate or specific environmental phenomena (Saucedo and Southgate, 2008).

In this regard, the development of the gonad is an energy-dependent process and the mobilization of nutrients from ingested food into the gonads is essential for the formation of gametes, a process that has been studied in detail in some species (Sastry, 1968, 1970, 1975, Sastry and Blake, 1971; Gimazane 1971; Bayne, 1975, 1976; Besnard et al. 1989; Napolitano and Ackman 1992; Pazos et al., 1997, Galap et al., 1997, Palacios et al., 2005).

In addition, numerous studies have shown a direct relationship between food availability and gonadal development in several marine bivalves, including *Aequipecten irradians* (Sastry, 1968), *Nodipecten subnodosus* (Palacios et al., 2005) and *Pecten maximus* (Pazos et al., 1997a).

In contrast, the species with two annual reproductive half cycles and that show possible conservative tactics in at least the second reproductive half cycle include: A. *seminuda, L. scabra, M. squamosus, N. nodosus* and *P. perna*. Among those considered to show a continuous reproduction with gametogenesis and gamete maturation during the period of low food availability (August to November) are: *A. notabilis, A. zebra, P. imbricata, C. rhizophorae, P. viridis, T. mactroides and D. denticulatus*.

All these aforementioned species have a tropical distribution.

A conservative reproductive strategy was also observed in *Atrina seminuda* populations in Argentina, where production of gametes occurs during the fall when food availability is low and appears to be supported by the energy reserves of the adductor muscle (Soria et al., 2002).

For example, when the temperature and the concentration of seston decrease in winter, some species follow a conservative reproductive strategy using energy reserves stored previously in somatic tissues. This occurs in *Pinctada fucata* from India (Desai et al., 1979), *P. mazatlanica* and *Pteria sterna* from *Bahia de La Paz*, Southern Baja California, Mexico (Saucedo and Southgate, 2008) and in the cockle *Cerastoderma edule* from Northern Spain (Navarro et al. 1989).

The coastal species must rely in part on energy reserves accumulated in their tissues to cope with the costly reproductive process under conditions of low food availability and high temperatures (which theoretically increases energy expenditure due to an increase in metabolic rate).

An alternative energy source to phytoplankton would be organic detritus from resuspension of sediments. Most marine bivalves are filter feeders and their food includes phytoplankton, bacteria, flagellates, dissolved organic matter and aggregates organic (Jorgensen, 1966). The detritus has been suggested as an important food source for bivalve filter feeders such as *Aequipecten irradians* (Kirby-Smith, 1976) and *Macoma balthica* (Newell, 1965). This was further highlighted by Freites et al. (2003) in the scallop *N. nodosus*, who showed that particulate organic matter of sedimentary origin played an important role as a source of energy to meet their metabolic and reproductive activities during the period of upwelling relaxation (low primary productivity and high water temperatures).

The information provided in this work demonstrates the adaptation to the environment of many species of bivalve distributed in an ecoregion with two annual periods or seasons that are quantitatively and qualitatively different in terms of food availability (energy). In particular, those species which have two reproductive types or tactics (opportunistic and conservative) can withstand such environmental changes to ensure their reproduction. This reveals a physiological flexibility to maximize the resources on offer and ensure the conservation of the species and its presence in marine environments.

## References

Acosta, V., Lodeiros, C. and Freites, L. 2010. Componentes bioquímicos de los tejidos de *Perna perna* y *Perna viridis* (L, 1758) (Bivalvia: Mytilidae), en relación al crecimiento en condiciones de cultivo suspendido. *Lat. Am. J. Aquat. Res.*, 38 (1): 37-46.

Alfaro, C. A., Jeffs, A. G. and Hooker, S. H. 2001. Reproductive behavior of the green-lipped mussel, *Perna canaliculatus*, in Northern New Zealand. *Bull. Mar. Sc.*, 69: 1095-1108.

Álvarez, M. 1992. Estudio ecológico de tres comunidades en la costa norte del estado Sucre, Venezuela. *Tesis de Pregarado*, Universidad de Oriente, Cumaná, Venezuela. 109 pp.

Andrews, J. 1979. Pelecypoda: Ostreidae. En: Giese, A. C. and Pearse, J. S. (Eds.). *Reproduction of Marine Invertebrates*. Academic Press. New York. pp. 293-363.

Angel-Pérez, C., Serrano-Guzmán, S. J., Ahumada-Sempoal, M. A. 2007. Ciclo reproductivo del molusco *Atrina maura* (Pterioidea: Pinnidae) en un sistema lagunar costero, al sur del Pacífico tropical mexicano. *Rev. Biol. Trop.* 55: 839-852.

Ansell, A. D., Loosmore, F. A. and Lander, K. F. 1964. Studies on the hard-shell clam, *Venus mercenaria*, in British waters. II. Seasonal cycle in condition and Biochemical composition. *J. Appl. Ecol.*, 1: 83-95.

Arellano-Martínez, M., Ceballos-Vázquez, B. P., Villalejo-Fuerte, M., García-Domínguez, F., Elorduy-Garay, J. F., Esliman-Salgado, A., and Racotta, I. S. 2004. Reproduction of the lion's paw scallop Nodipecten subnodosus Sowerby, 1835 (Bivalvia: Pectinidae) from Laguna Ojo de Liebre, B.C.S., México. *J. Shellfish Res.* 23 (3): 723–729.

Arrieche, D., Liceo, B., García, N., Lodeiros, C., and Prieto, A. 2002. Índice de condición, gonádico y de rendimiento del mejillón marrón *Perna perna*

(Bivalvia: Mylilidae), del Morro de Guarapo, Venezuela. *Inter-ciencia,* 27 (11): 613-619.

Arsenault, D. J. and Himmelman, J. H. 1998. Spawning of the Iceland scallop (*Chlamys islandica*, Muller, 1776) in the northern Gulf of St. Lawrence and its relationship to temperature and phytoplancton abundance. *Veliger*, 41:180-185.

Avendaño, M., Cantillánez, M., Le Pennec, M., and Thouzeau, G. 2008. Reproductive and larval cycle of the scallop *Argopecten purpuratus* (Ostreida: Pectinidae), during El Nino- La Nina events and normal condition in Antofagasta, Northen Chile. *Rev. Biol. Trop.*, 56: 121–132.

Barber, B. J. and Blake, N. J. 1991. Reproductive Physiology. E. Shumway, S. E. (Ed.). *Scallops: Biology, Ecology and Aquaculture*. Elsevier. Amsterdam. pp. 377-428.

Bayne, B. L. 1975. Reproduction of bivalve mollusks under environmental stress. In: Vernberg, F. J. (ed.) *Physiological ecology of estuarine organisms*. University South Carolina Press, Columbia, pp. 259-277.

Bayne, B. J. 1973. Physiological changes in *Mytilus edulis* induced by temperature and nutritive stress. *J. Mar. Biol. Assoc. UK* 53: 39–58.

Bayne, B. L. 1976. Aspects of reproduction in bivalve mollusks. In: M. Wiley, Ed., *Estuarine Processes. Uses, stresses and adaptation to the estuary*. Vol. 1. Academic Press, New York. pp. 432-448.

Bayne, B. L. and Newell, R. C. 1983. Physiological Energetic of Marine Mollusks. En: Saleuddin A. S. M. and Wilbur, K. M. (Eds.). *The Mollusca*. Academic Press, New York. Vol. 4. pp. 491-498.

Beninger, P. G. and Le Pennec, M. 1997. Reproductive characteristics of a primitive bivalve from a deep-sea reducing environment: giant gametes and their significance in *Acharax alinae* (Cryptodonta: Solemyidae). *Mar. Ecol. Prog. Ser.*, 157: 195–206.

Beninger, P. G. and Le Pennec, M. 2006. Structure and function in scallops. En: Scallop: Biology, Ecology and aquaculture. Shumway, S. E. and Parsons, G. J. (Eds.). *Developments in aquaculture and Fisheries Sciences*. 123-211 pp. Elsevier, Amsterdam, The Netherlands.

Berg C. J. and Alatalo, P. 1985. Biology of the tropical bivalve *Asaphis defl orata* (Linné, 1758). *Bull. Mar. Sci.*, 37: 827-838.

Besnard, J. Y., Lubet, P. and Nouvelot, A. 1989. Seasonal variations of the fatty acid content of the neutral lipids and phospholipids in the female gonad of *Pecten maximus* L. *Comp. Biochem. Physiol.*, 93B: 21–26.

Beukema, J. J. and De Bruin, W. 1977. Seasonal changes in dry weight and biochemical composition of the soft parts of the tellinid bivalve *Macoma balthica* in the Dutch Wadden Sea. *Neth. J. Sea. Res.*, 11: 42-55.

Brea, J. M. 1986. Variaciones estaciónales en la composici6n bioquímica de *Pecten ziczac* (Linn. 1758) en relación al metabolismo energético, reproducci6n y crecimiento. *Tesis del Departamento de Biología, Escuela de Ciencias,* Universidad de Oriente, Cumana, Venezuela.

Cantillánez, M., Avendaño, M. and Thouzeau, G. 2005. Reproductive cycle of *Argopecten purpuratus* (Bivalvia: Pectinidae) in La Rinconada marine reserve (Antofagasta, Chile): response to environmental effects of El Niño and La Niña. *Aquaculture*, 246: 181–195.

Cruz-Soto, Rafael Angel. 1987. Tamaño y madurez sexual en *Anadara grandis* (Pelecypoda: Arcidae). *Brenesia*. 27: 9-12.

Dare, P. J. and Edwards, D. B. 1975. Seasonal changes in flesh weight and biochemical composition of mussels (*Mytilus edulis* L.) in the Conway estuary, North Wales. *J. Exp. Mar. Biol. Ecol.*, 18: 89-97.

Desay, K., G. Hirani and D. Nimavat. 1979. Studies on the pearl oyster *Pinctada fucata* (Gould): Seasonal biochemical changes. *Ind. J. Mar. Sci.*, 8: 49-50.

Eble, A. E. 2001. Anatomy and Histology of *Mercenaria mercenaria*. En: *Biology of the Hard Clam*, Volume 31. 117-216 pp. Developments in Aquaculture and Fisheries Science. Kraeuter, J. N. and Castagna, M. (Eds.). Elsevier. The Netherlands.

Fournier, M. L. 1992. The reproductive biology of the tropical rocky oyster *Ostrea iridescens* (Bivalvia: Ostreidae) on the Pacific coast of Costa Rica. *Aquaculture*, 101: 371–378.

Freites, L., Lodeiros, C., Narváez, N., Estrella, G., and Babarro, J. M. F. 2003. Growth and survival of the scallop *Lyropecten* (=*Nodipecten*) *nodosus* (L., 1758) in suspended culture in the Cariaco Gulf (Venezuela), during a non-upwelling period. *Aquacult. Res.*, 34: 709-718.

Freites, L., Montero, L., Arrieche, D., Babarro, P., Saucedo, J. M. F. and Cordova, C. and García, N. 2010a. Influence of the environmental factors on the reproductive cycle of the tropical bivalve *Anadara notabilis* (Röding, 1798*). J. Shellfish Res.*, 29 (1): 69-75.

Freites, L., Cordova, C., Arrieche, D., Montero, L., García, N., and Himmelman, J. H. 2010b. Reproductive cycle of the penshell *Atrina seminuda* (Mollusca: Bivalvia) in tropical waters. *Bull. Mar. Scienc*. 86 (4): 785-801.

Frenkiel, L., Gros, O. and Moueza, M. 1977. Storage tissue and reproductive strategy in *Lucina pectinata* (Gmelin), a tropical lucinid bivalve adapted to a reducing sulfuric, mangrove environment. *Invertebr. Rep. Dev.*, 31: 199–210.

Gabbottt, P. A. and Bayne, B. L. 1973. Biochemical effects of temperature and nutritive stress on *Mytilus edulis* (L.) *J. Mar. Biol. Ass. UK,* 53: 269-286.

Galap, C., Leboulenger, F. and Grillot, J. P. 1997. Seasonal variations in biochemical constituents during the reproductive cycle of the female dog cockle *Glycymeris glycymeris*. *Mar. Biol.*, 179: 625-634.

Galtsoff, P. S. 1964. The American Oyster *Crassostrea virgínica* (Gmelin). *Fish. Bull.,* 64: 1-480.

García, N., C. Lodeiros, D. Arrieche, A. Prieto, L. Freites and J. Himmelman. 2007. Relación del tejido somático y los factores ambientales en el ciclo reproductivo de la vieira (*Nodipecten nodosus*) (L., 1758). *Bol. Centro Invest. Biol. L.U.Z.*, 41: 292-308.

Giese, A. C. and Pearse, J. S. 1974. *Reproduction of Marine Invertebrates*: Academic Press. 546 pp.

Giles, W. Ciclo reproductivo e índice condición de la especie *Anadara notabilis* (Roding, 1798) en playa Tocuchare, Estado Sucre, Venezuela. *Tesis de grado* (Licenciatura en Biología, Mención Biología Marina). Universidad de Oriente, Núcleo de Sucre, Escuela de Ciencias, Departamento de Biología, 1984.

Gimazane, J. P. 1971. Etude expérimentale de l'action de quelques facteurs externes sur la reprise de l'activité génitale de la coque, *Cerastoderma edule* L., Mollusque bivalve. *C. R. Soc. Biol.*, 166: 587-589.

Gómez-Robles, E. and Saucedo, P. 2009. Evaluation of quality indices of the gonad and somatic tisúes involved in reproduction of the pearl oyster *Pinctada mazatlanica* with histochemistry and digital image analysis. *J. Shellfish res*. 28: 329-335.

Haag, W. R. and Garton, D. W. 1992. Synchronous spawning in a recently established population of the zebra mussel, *Dreissena polymorpha,* in western Lake Erie, US. *Hydrobiology*, 234: 103-110.

Jaramillo, R., Winter, J., Valencia, J., and Rivera, A. 1993. Gametogenic cycle of the chiloé scallop (*Chlamys amandi*). *J. Shellfish Res*., 12: 165-172.

Jørgensen, C. B. 1966. *Biology of suspension feeding*. Pergamon, New York.

Joseph, M. M. and Madhyastha, M. N. 1984. Annual reproductive cycle and sexuality of the oyster *Crassostrea madrasensis* (Preston). *Aquaculture*, 40: 223–231.

Kang, C. K., Park, M. S., Lee, P. Y., Choi, W. J., and Lee, W. C. 2000. Seasonal variations in condition, reproductive activity, and biochemical composition of the Pacific oyster, *Crassostrea gigas* (Thunberg), in suspended culture in two coastal bays of Korea. *J. Shellfish Res.* 19: 771-778.

Kirby-Smith, W. 1976. The detritus problem and the feeding and digestion of an estuarine organism. In: *Estuarine processes,* Wiley, M. (Ed.). Academic Press, New York. Vol. 1, 469-479 pp.

Lasiak, T. 1986. The reproductive cycles of the intertidal bivalves *Crassostrea cucullata* (Born, 1778) and *Perna perna* (Linnaeus, 1758) from the Transkei coast, Southern Africa, *Veliger*, 29: 226–230.

Lista, M., Lodeiros, C., Prieto. A., Himmelman, J. H., Castañeda, J., García, N., and Velásquez, C. 2006. Relation of seasonal changes in the mass of the gonad and somatic tissues of the zebra ark shell *Arca zebra* to environmental factors. *J. Shellfish Res.*, 25:969-973.

Lodeiros, C. J. and Himmelman, J. H. 1994. Relations among environmental conditions and growth in the tropical scallop *Euvola (Pecten) ziczac* (L.) in suspended culture in the Golfo de Cariaco, Venezuela. *Aquaculture,* 119:345-358.

Lodeiros, C. J. and Himmelman, J. H. 1999. Reproductive cycle of the bivalve Lima scabra (Pterioida: Limidae) and its association with environmental conditions. *Rev. Biol. Trop.* 47(3): 411-418.

Lodeiros, C., Marín, B. and Prieto, A. 1999. *Catálogo de moluscos marinos de las costas nororientales de Venezuela: Clase Bivalvia.* Edición APUDONS. 109 p.

Lodeiros, C. J. and Himmelman, J. H. 2000. Identification of factors affecting growth and survival of the tropical scallop *Euvola* (*Pecten*) *ziczac* in suspended culture in the Golfo de Cariaco, Venezuela. *Aquaculture*, 182: 91-114.

Lodeiros, C. J., Alio, J. and Marcano, J. 2006. Actividad extractiva de moluscos en Venezuela. En: J. Fernández, M. Rey and A. Guerra (eds.). Memorias del VIII Foro sobre los Recursos Marinos y Acuicultura de las Rías Gallegas. *Xunta de Galicia-Universidad de Santiago*. Págs. 353–360. El Grove, España.

Luna-González, A., Cáceres-Martínez, C., Zuñiga-Paheco, C., López-López, S., and Ceballos-Vásquez, B. P. 2000. Reproductive cycle of *Argopecten ventricosus* (Sowerby II. 1842). (Bivalvia-Pectinidae) in the Rada of Puerto Pichilingue, B. C. S., Mexico and its relation to temperature, salinity and quality of food. *J. Shellfish Res.* 19: 107-112.

MacDonald, B. A. and Thompson, R. J. 1985a. Influence of temperature and food availability on the ecological energetic of the giant scallop *Placopecten magellanicus*. I. Growth rates of shell and somatic tissue. *Mar. Ecol. Prog. Ser.*, 25: 279-294.

MacDonald, B. A. and Thompson, R. J. 1985b. Influence of temperature and food availability on the ecological energetic of the giant scallop *Placopecten magellanicus*. II. Reproductive output and total production. *Mar. Ecol. Prog. Ser.*, 25: 295-303.

Mackie, G. L. 1979. Growth dynamics in natural populations of Sphaeriidae clams (*Sphaerium, Musculium, Pisidium*). *Can. J. Zool.*, 57: 441-456.

Mackie, G. L. 1984. Bivalves. En: *The Mollusca: Reproduction*. Tompa, A. S., Verdonk, N. H. and Van den Biggelar, J. A. M. (Eds.). Academic Press, London, Vol. 7. pp. 351-418.

Malachowsky, M. 1988. The reproductive cycle of the rock scallop *Hinnites giganteus* (Grey) in Humboldt Bay, California. *J. Shellfish Res.*, 7: 241-348.

Mandelli, E. and E. Ferráz-Reyes. 1982. Primary production and phyto-plankton dynamics in a tropical inlet, Gulf of Cariaco, Venezuela. *Inter. Revue. Ges. Hydrobiol.*, 57: 85-95.

Marcano, V. 1984. Aspectos biológicos de la reproducción en la ostra perla *Pinctada imbricata* (Röding, 1798) (Mollusca: Bivalvia) de Punta Las Cabeceras, Isla de Cubagua, Venezuela. *Tesis de Grado*, Departamento de Biología, Universidad de Oriente.

Marcano, M. 2004. Histología gonadal de *Perna viridis* L 1758 (Bivalvia: Mytilidae) del morro de Guarapo, costa norte del estado Sucre, Venezuela. *Tesis. Universidad de Oriente*. Cumaná Venezuela. 54 pp.

Miloslavich, P. and Klein, E. 2008. Ecorregiones marinas del Caribe venezolano. p: 16-19. In: Klein, E. (Ed.). Prioridades de PDVSA en la conservación de la biodiversidad en el Caribe venezolano. Petróleos de Venezuela, S.A. Universidad Simón Bolívar. *The Nature Conservancy*. Caracas, Venezuela. 72 p.

Mora, J. 1985. Distribución por tallas, ciclo gonádico e índice de engorde de la pepitona *Arca zebra,* Boca de Rio, Isla de Margarita. *Tesis de Pregrado*, Universidad de Oriente, Cumaná, Venezuela. 95 pp.

Morriconi, E., Lomovasky, B. J., Calvo, J., and Brey, T. 2002. The reproductive cycle of *Eurhomalea exalbida* (Chemnitz, 1795) (Bivalvia: Veneridae) in Ushuaia Bay (54° 50'S), Beagle Channel (Argentina). *Invertebr. Repr. Dev.*, 42: 61-68.

Muller-Karger, F. and Varela, R. 1988: Variabilidad de la biomasa del fitoplancton en aguas superficiales del mar Caribe: una perspectiva desde el espacio con el CZCS. *Fund. La Salle Contrib.*, 171: 1-23.

Napolitano, G. E. and Ackman, R. G. 1992 Anatomical distributions and temporal variations of lipid classes in sea scallops *Placopecten magellanicus* (Gmelin) from Georges Bank (Nova-Scotia). *Comp. Biochem. Physiol.* 103B, 645-650.

Navarro, E., Iglesias, J. I. P. and Larrañaga, A. 1989. Interannual variation in the reproductive cycle and biochemical composition of the cockle *Cerastoderma edule* from Mundaca Estuary (Biscay, Noth Spain). *Mar. Biol.,* 101: 503-511.

Newell, R. C. 1965. The role of detritus in the nutrition of two marine deposit feeders, the prosobranch *Hydrobia ulvae* and the bivalve *Macoma balthica*. *Proc. Zool. Soc. London*, 144: 25-45.

Okuda, T., Benitez, J., Sellier, J., Fukuoka, J., and Gamboa, B. (1974): Revisión de los datos oceanográficos en el mar Caribe Suroriental, especialmente el margen continental de Venezuela, *Cuadernos Azules,* 15: 1-179.

Pazos, A. J., Roman, G., Acosta, C. P., Abad, M., and Sánchez, J. L. 1997b. Influence of the gametogenic cycle on the biochemical composition of the ovary of the great scallop. Especial Issue of the Scallop Workshop. *Aquac. Intern.*, 4(3): 201-213.

Pazos, A. J., Román, G., Acosta, C. P., Abad, M., and Sánchez, J. L. 1997c. Seasonal changes in condition and biochemical composition of the scallop *Pecten maximus* L. from suspended culture in the Ria de Arousa (Galicia, N.W. Spain) in relation to environmental conditions. *J. Exp. Mar. Biol. Ecol.*, 211: 169-193.

Pieters, H., Kluytmans, J. H., Zurburg, W., and Zandee, D. I. 1979. The influence of seasonal changes on energy metabolism in *Mytilus edulis* L. - I. Growth rate and biochemical composition in relation to environmental parameters and spawning. In: *Cyclic Phenomena in Marine Plants and animals*. Naylor, E. and Hartnoll, R. G. (Edit.). Pergamon Press, New York. Pp. 285-292.

Pirela-Ochoa, E., Troccoli, L. and Hernández-Ávila, I. 2008. Hidrografía y cambios en la comunidad del microfitoplancton de la Bahía de Charagato, Isla de Cubagua, Venezuela. *Bol. Inst. Oceanogr. Univ. Oriente*, Venezuela 47, 3–15.

Prieto, A. 1980. Contribución a la ecología de *Tivela mactroides* (Bor, 1778): *Aspactos reproductivos. Bol. Inst. Oceanog.* Univ. Sao Paulo. 29(2): 323-328.

Prieto, A. 1983. Ecología de *Tivela mactroides* (Born, 1778) (Mollusca: Bivalvia) en la playa Güiria. *Bol. Inst. Oceanogr.* Univ. Oriente. 22 (1-2):7-19.

Prieto, A. S., Pereira, R. and Manrique, R. 1985. Producción secundaria del mejillón *Modiolus squamosus* (Beauperthuy, 1967) en Tocuchare, Golfo de Cariaco. Venezuela. *Acta Cient. Venezolana*, 36: 258-264.

Prieto, A., J. Marcano, L. Villegas y C. Lodeiros. 2008. Estructura poblacional de la almeja, *Asaphis deflorata* en la localidad de Caurantica, Golfo de Paria, estado Sucre, Venezuela. *Zoot. Trop.* 26 (1): 55-62.

Racotta, I., Ramírez, J. L., Avila, S., and Ibarra, A. 1998. Biochemical composition of gonad and muscle in the catarina scallop *Argopecten ventricosus*, after reproductive conditioning under feeding systems. *Aquaculture*, 163: 111-122.

Racotta, I., Ramírez, J., Ibarra, A., Rodríguez-Jaramillo, M., Carreño, D., Palacios, E., 2003. Growth and gametogenesis in the lion's paw scallop *Nodipecten (Lyropecten) subnodosus*. *Aquaculture*, 217: 335-349.

Ruffini, E. 1984. Desarrollo larval experimental de la ostra perla *Pinctada imbricata* (Röding, 1798) (Mollusca: Bivalvia) y algunas observaciones sobre su reproducción en el banco natural de Punta Las Cabeceras, Isla de Cubagua, Venezuela. *Tesis de Grado.* Universidad de Oriente, Cumaná, Venezuela. 51 pp.

Ruiz, E., Cabrera, E., Cruz, P. R., and Palácios, J. A. 1998. Crecimiento y ciclo reproductivo de *Polymesoda radiata* en Costa Rica. *Rev. Biol. Trop.* 46: 643-649.

Saint-Aubyn, M., A. Prieto and L. Ruiz. 1992. Producción específica de una población del bivalvo *Arca zebra* (Swainson, 1883), en la costa nororiental del estado Sucre, Venezuela. *Act. Cient. Venolana.* 50:15–23.

Sastry, A. N. 1968. Relationship among food, temperature and gonadal development of the bay scallop, *Aequipecten irradians* Lamark. *Physiol. Zool.*, 41: 44-53.

Sastry, A. N. 1970. Reproductive physiological variation in latitudinaly separated populations of the bay scallop *Aequipecten irradians* Lamarck. *Biol. Bull.*, 138: 56-65.

Sastry, A. N. 1975. Physiology and ecology of reproduction in marine invertebrates. In: *Physiological Ecology of Estuarine Organisms*,

Vernberg, F. J. (Ed.) University of South Carolina Press, Columbia. 279-299 pp.

Sastry, A. N. 1979. Pelecypoda (excluding Ostreidae). En: Giese, A. C. and Pearse, J. S. (Eds.). *Reproduction of Marine Invertebrates*. Academic Press. New York. pp. 113-292.

Sastry, A. N. and Blake, N. J. 1971. Regulation of gonad development in the bay scallop, *Aequipecten irradians* Lamarck. *Biol. Bull.* (Wood Hole, Mass.) 140: 274-283.

Saucedo, P. and Southgate, P. 2008. Reproduction, development and growth. In: *The Pearl Oyster* Southgate, P. and Lucas, J. (Eds.) Elsevier, 131-186.

Seed, 1976. Ecology. En: Bayne, B. L. (Ed.). *Marine mussels: their Ecology and Physiology*. Cambridge Univ. Press. Cambridge. Pp. 13-65.

Shafee, M. S. 1989. Reproduction of *Perna picta* (Mollusca: bivalvia) from the Atlantic coast of Morocco, *Mar. Ecol. Prog. Ser.* 53: 235–245.

Soria, R., Pascual, M. and Fernández, V. 2002. Reproductive cycle of the cholga paleta *Atrina seminuda* (Lamarck, 1819) from Northern Patagonia, Argentina. *J. Shellfish Res.*, 21: 479-488.

Taylor, A. and Venn, T. 1979. Seasonal variation in weight and biochemical composition of the tissues of the queen scallop *Chlamys opercularis*, from the Clyde sea. *J. Exp. Mar. Biol.*, 59: 605-621.

Thompson, J. T., Newell, R. I. E., Kennedy, V. S., and Mann, R. 1996. Reproductive processes and early development. En: Kennedy, V. S., Newell, R. I. E. and Eble, A. F. (Eds). The eastern Oyster *Crassostrea virginica*. Maryland. *Maryland Sea Grant Book*. pp. 335-370.

Varela, R., Carvajal, F. and Muller-Karger, F. 2003. El fitoplancton en la plataforma nororiental de Venezuela. En: La sardina (*Sardinella aurita*). Su medio ambiente y explotación en el Oriente de Venezuela. Fréon, P. and Mendoza, J. (Eds). *IRD Éditions. Collection Colloques et sémina-ries*. Pp. 263-294.

Vélez, A. 1977. Ciclo anual de reproducción del ostión *Crassostrea rhizophorae* (Guilding), de bahía Mochima. *Bol. Inst. Oceanog. Univ. Oriente*, 16: 87-98.

Vélez, A. 1985. Reproductive biology of the tropical clam *Donax denticulatus* in eastern Venezuela. *Carib. J. Sci.* 21: 125-136.

Vélez, A. and Martínez, R. 1967. Reproducción y desarrollo larval experimental del mejillón comestible de Venezuela *Perna perna* (Linnaeus, 1758). *Bol. Inst. Oceanogr. Univ. Oriente*, 6: 266-285.

Veléz, A. 1991. Biology and culture of the Caribbean or Mangrove oyster *Crassostrea rhizophorae* (Guilding), in the Caribbean and South

American. In: *Estuarine and Marine Bivalve Mollusk Culture*, Menzel, W. (Ed.) 118-123 pp.

Vélez, A., Sotillo, F. and Pérez, J. 1987. Variación estacional de la composición química de los pectínidos *Euvola* (*Pecten*) *ziczac* y *Lyropecten nodosus*. *Bol. Inst. Oceanogr. Univ. Oriente*, 26: 67-72.

Vélez, A., Alifa, E. and Freites, L. 1993. "Inducción a la reproducción en la vieira *Pecten ziczac* (Mollusca: Bivalvia). Maduración y Desove. *Caribbean J. Sci.*, 29 (3-4): 209-213.

Villalba, A. 1995. Gametogenic cycle of culture mussel, *Mytilus galloprovincialis*, in the bays of Galicia (N. W. Spain). *Aquaculture*, 130: 269-277.

Vite-García, M. N. 2005. Almacenamiento y utilización de reservas energéticas en relación con la reproducción de las ostras perleras *Pteria sterna* y *Pinctada mazatlanica* (Bivalvia: Pteriidae) *Tesis de Maestría*, Centro de Investigaciones Biológicas del Noroeste. La Paz, México: 95 p.

Vite-Garcıía, M. N. and P. E. Saucedo. 2008. Energy storage and allocation during reproduction of pearl oyster *Pteria sterna* (Gould, 1851) at Bahía de La Paz, Baja California Sur, México. *J. Shellfish Res.* 27: 375–383.

White, K. M. 1937. *Mytilus*. Mem. Liverpool. *Mar. Biol. Comm.* 37: 1-117.

Wootton, R. J. 1984. Introduction: Tactics and strategies in fish reproduction. En: *Fish reproduction: Strategies and tactics*. Potts, G. W. and Wootton, R. J. (Eds.). Academic Press, 1-12 pp.

Zandee, D. I., Holwerda, D. A. and de Zwaan, A. 1980. Energy metabolism in bivalves and cephalopods. In: *Animals and environmental Fitness*. Gilles, R. (edit.). Vol. 1. pp. 185-206.

In: Spawning
Editor: Erick R. Baqueiro Cárdenas
ISBN: 978-1-63117-655-5


*Chapter 5*

# REVIEW OF REPRODUCTIVE TACTICS OF BIVALVE AND GASTROPOD MOLLUSKS FROM MEXICO

***Erick R. Baqueiro Cárdenas*[1] *and Dalila Aldana Aranda*[2]**
[1]Instituto Tecnológico Superior de Champotón, Champotón, Mexico
[2]Centro de Investigación y de Estudios Avanzados, del Instituto Politécnico Nacional, Mérida, México

## ABSTRACT

The reproductive cycles are analyzed with seven Bivalves and six gastropods to determine the variability and similarities of their reproductive cycle under diverse environmental conditions. Some species were studied at the same localities at different times and at different localities, also different species were studied simultaneously at the same locality to understand adaptive tactics to attenuate interspecific competition. Species inhabiting the same locality differ in their spawning cycle and periods of maximum activity. This study showed two gametogenic strategies as a response to the environment: populations of fast gametogenesis during a short lapse of time, and populations with a continuous gametogenesis through most of the year. The variations on the reproductive cycle of a species among different localities can be associated with environmental levels of instability. Predation and competition induce massive spawning and reduction of the spawning periods. Species inhabiting the same locality and similar habitats differ in their spawning cycle based on periods of maximum activity. Variations of one species among localities can be associated with environmental instability or variations in critical parameters with a tendency to optimize reproduction through one of several alternatives: fast gametogenesis with accumulation of mature gametes, slow constant gametogenesis with limited or no accumulation of mature gametes, constant asynchronic spawning, and synchronic spawning from limited to extended periods of time.

**Keywords**: Gonad development, environmental factors, Mollusks, reproductive system

## INTRODUCTION

Reproductive tactics are the result of genetic and environmental interactions and a response to life strategies and habitat where populations live (Mackie, 1984). Species are classified in two groups, given the duration of their reproductive cycle: (1) taquitictic, with short reproductive periods, and (2) braditictic, with extended reproductive periods. Species with a wide geographical distribution show an ample variation of reproductive strategies in both duration and intensity (Bricej and Malouf, 1980; Kennedy and Kratz, 1982; Knaub and Eversole, 1988). This has been linked to latitude, but with more precision to temperature and food availability (Bayne, 1978; Sastry, 1979; Webber, 1977; Fretter, 1984; Mackie, 1985), which have been experimentally proven by conditioning in the laboratory organisms of one species from different populations (Loosanoff and Davis1963; Bayne, 1978; Ino, 1970; Lubet and Choquet, 1971; Hines, 1979; Castagna and Kraeuter, 1981). As for maturity and reserve storage, populations of the same species may mature during the fall and store their mature gametes through winter to spawn in spring, as shown for *Spisula solidissima* from Georgia, U.S.A. (Kanti et al., 1993), or present a long, undifferentiated stage of lipid and carbohydrate storage during winter, withholding gametogenic activity until spring, as in Prince Edwards Island, Canada (Shephton, 1987). Likewise, the same population may show differences in its reproductive activity from one year to another, from a short synchronic spawn, a long, partially asynchronic spawn, or several partial spawns (Bricelj and Malouf, 1980; Baqueiro, 1981; Jaramillo et al., 1993).

In Mexico, over 80 species of bivalves and gastropods are exploited commercially (Baqueiro and Castagna, 1988). They range from the temperate waters of the western coasts of Baja California to the eastern Caribbean coast of Yucatán with very dry to wet tropical weather patterns. They also occur in coastal hypersaline and hyposaline lagoons and in oceanic environments that favour the expression of a variety of reproductive tactics. It is the scope of this chapter to present variations of the reproductive cycle of species living under different climatic conditions and of different species living in the same locality.

## Material and Methods

Chart 1 presents the seven bivalve and six gastropod species studied from eleven different localities. Over the course of over 30 years of fisheries biology studies of commercially exploited populations, at least twenty specimens of each population were collected monthly to cover an annual cycle. The shell length and width were measured for every animal and the whole organism and soft parts were weighted after extraction at the laboratory. A section of the visceral mass and gonad were fixed with Bouin solution or neutral formalin (Luna, 1968). All samples were rinsed for 12 hours with running tap water and preserved in alcohol at 70% with glycerin at 0.1%. Sections of 1 $cm^3$ were dehydrated with alcohol, cleared with Xylen, and included in 53°C - 56°C melting point paraffin. Sections were cut 7 µm and 12 µm for males and females respectively with a rotation microtome. All species were stained with Hematoxiline-eosine (Luna, 1968).

The practical classification of Lucas (1965) was used to define the reproductive cycle in five stages: Stage I Rest—the sex may not be identified from the microscopic section; Stage II Gametogenesis—active cell division, mature gametes may or may not be present; Stage III Mature—dominance of mature gametes, although some gametogenesis may be present; Stage IV Spawn—partially emptied follicles; Stage V Post-spawn—follicles are partially or totally emptied and broken, eggs and sperms are being reabsorbed, phagocytes are present.

## Sampling Sites

### West Coast of the Baja California

The west coast of the Baja California peninsula has a desert climate with winter rains of less than 70 mm $yr^{-1}$ and a mean annual temperature of

20.7°C, mean maximum of 27.1°C during September, and mean minimum of 15.3°C during February.

**Chart 1. Species, locality and habitat of populations used in this analysis**

| Species | Locality | Habitat |
|---|---|---|
| *Mytella strigata* | Chautengo and Nuxco lagoons, Guerrero. | Intertidal, muddy bottom |
| *Chione undatella* | La Paz, B.C.S | Intertidal, sandy bottom |
| *Anadara tuberculosa* | La Paz, B.C.S | Intertidal, mangrove mud |
| *Argopecten ventricosus* | La Paz, Concepción Bay, Ojo de Liebre lagoons, B.C.S | Intertidal, Thallassia |
| *Dosinia ponderosa* | Zihuatanejo, Guerrero | Sublitoral, sandy bottom |
| *Megapitaria squalida* | Zihuatanejo, Guerrero | Sublitoral, sandy bottom |
| *M. aurantiaca* | Zihuatanejo, Guerrero | Sublitoral, sandy bottom |
| *Hexaplex erythrostomus* | Conceptions bay, Gulf of California. | Sublittoral, protected |
| *Strombus gracilior* | Conceptions bay, Gulf of California. | Sublittoral, protected |
| *Strombus pugilis* | Seyba playa, Gulf of Mexico | Sublittoral, exposed |
| *Melongena corona* | Campeche bay, Gulf of Mexico | Sublittoral, exposed |
| *Fasciolaria tulipa* | Campeche bay, Gulf of Mexico | Sublittoral, exposed |
| *Strombus gigas* | Chinchorro reef, Caribbean sea, Alacranes reef, Gulf of México | Sublittoral, protected coral reef lagoon |

The desert climate on the eastern coast is characterized by rain in summer and winter, with an average rainfall of 200 mm annually. The mean annual temperature is 24°C, with a monthly maximum mean of 29.6°C and monthly minimum mean of 18°C (Garcia, 1981).

## Guerrero Negro Lagoon

Ojo de Liebre and Guerrero Negro lagoons form one system with separate mouths opening into Vizcaino Bay (Figure 1A). Guerrero Negro lagoon is located between 27°55′–28°05′N and 114°03′–114°10′W, with a total surface of 160 $km^2$ and a wide mouth opening into Vizcaino Bay. Maximum tidal range is 2.61 m, generating currents of up to 50 cm $s^{-1}$ (Phleger, 1965) that form channels 7 m in depth between the sea grass beds and sandy intertidal banks. The mean depth of the lagoons is 1.5 m.

Average ambient temperature fluctuates from 12.3°C in December to 20.4°C in August. The hydrologic balance (difference between rainfall and evaporation) is negative through the year, with an annual rainfall of 144.4 mm and a total evaporation of 2339.80 mm.

The maximum evaporation occurs during July and minimum during December–January, yielding a hypersaline environment with the saline gradient increasing from the mouth inwards.

## Concepción Bay

Concepción Bay is located at 26°31′–26°53′N and 111°41′–111°56′W. It is 43 km long and 11 km wide, parallel to the Gulf of California, with its mouth to the north (Figure 1B).

The mean depth is 22 m with a maximum of 37 m along the central part. Ambient temperature varies from 15.1°C during January to 31.1°C in July and there is a total annual rainfall of 113.1 mm. During the sampling period, evaporation exceeded precipitation through the year with significant rain only during January and September. Water temperature fluctuated from a minimum of 20°C during January to 34°C in October, with salinity reaching its highest value of 38‰ during June 1980 and the lowest of 35‰ during June 1979.

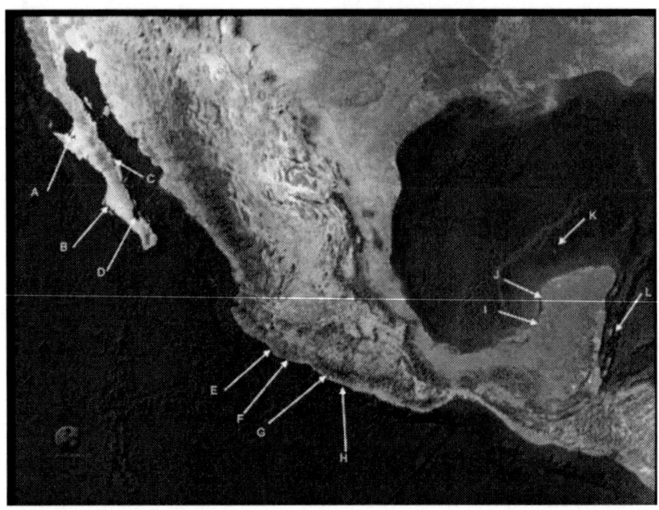


Figure 1. Sampling sites for the species included in this review: A) Guerrero Negro, B) Magdalena Bay, C) Conception Bay, D) La Paz Bay, Baja California Sur; E) Rio Balsas, F) Zihuatanejo, G) Chautengo Lagoon, H) Nuzco Lagoon, Guerrero; I) Seibaplaya, J) Selestun, Campeche; K) Alacranes reef, Yucatan; L) Chinchorro reef, Quintana Roo.

## Ensenada de La Paz

Ensenada de La Paz is located at 24°15'N and 110°21'W; it has an approximate surface area of 50 $km^2$ (Figure 1C). The depth fluctuates from 0.3 to 6 m. Tides are mixed with a daily variation of 1.4 to 2.5 m. The bottom consists of sand and clay with patches of *Thallassia* where *A. ventricosus* is found. A mean ambient temperature of 18°C is reached during January and a maximum of 29.6°C during August. During the sampling period, the lowest temperature was recorded during February and the highest during July. Salinity reached its highest level (38‰) during April and a minimum (34.5‰) during March. As in the previous three localities, evaporation exceeded precipitation, although at this locality rainfall was recorded for five months.

## Zihuatanejo

Zihuatanejo Bay is located in the north of the State of Guerrero on the Pacific coast of Mexico, at 17°37'N and 101°33'W (Figure 1D). The high-energy rocky coast has sandy pocket beaches influenced by the Balsas River (Lankford, 1974). Sediments are fine sand and silt with coarse sand around rocky reefs. The mean water temperature was 26.5°C with a maximum of 29.8°C and minimum of 23.2°C. Salinity was highest (33.87‰) during December and lowest (31.37‰) during August.

## Chautengo and Nuxco Lagoons

Both lagoons are located on the central portion of the Pacific coast of Guerrero state. Chautengo has an area of 33.5 km$^2$ and 600 m of coastal front and is connected by a narrow mouth to the ocean through most of the year (Figure 1F). The depth varies from 0.97 to 1.20 m. Sediments are mainly fine silt. The Nexpa and Copala rivers drain on the west end of the lagoon (Villarroel, 1975). Nuxco lagoon is located north of Acapulco at 17°11′–17°14′N and 100°47′–100°49′W. It is connected to the Pacific Ocean by a temporal mouth that opens during September. The Nuxco River drains on the northwest side (Figure 1G). The maximum depth is 3.5 m and the sediments are silt at the south and a silt-sand at the north. Temperatures in both lagoons varied from 28.5° to 33°C. Salinity, however, showed stronger variations during the year at Chautengo with a maximum of 41.46‰ and a minimum of 0.8‰, while at Nuxco the maximum was 24.24‰ and minimum 12.93‰.

## Campeche Bank

Campeche bank is located between 18° - 20° N and 90° - 91° W, on the Southwest Gulf of Mexico on the Yucatan peninsula (Figure 1H). The central and northern coasts of Campeche are part of the Yucatan peninsula of Karstic origin.

The central part is limited to the South by the Champoton River and to the North by the city of Campeche, presenting a rocky coast of low profile and sandy pocket beaches, with fine sediments from coastal mangroves and small coastal marshes. An ample tidal flat, with abundant fresh water upwelling, forms the northern part. Coastal vegetation is composed of mangrove swamps with muddy and sandy beaches (Lankford, 1974). Tides are semidiurnal (Graffé, 1990). The weather is hot and humid with summer rains, with 5.8% of them during fall; average temperature is 23° C; average rainfall is 1132 mm (Garcia, 1981).

## Chinchorros Reef

Chinchorro Reef is a false atoll located 16 miles off the coast from southern Quintana Roo, between 18°23'- 18°47' N and 87°14'- 87°27' W (Figure1I). It is oval 46 Km long, 19 Km in the widest part, with an area of 561 $Km^2$.

The depth inside the lagoon reef decreases from 12 m in the south region to 7 to 3 m in the central part to 2 m in the north. Chinchorro has four keys; the two small keys are known as Cayo Norte, the largest are Cayo Centro and Cayo Lobos, with the smallest located on the southern end. The weather is hot and humid with summer and winter rains, with average rainfall of 1249 mm; the average temperature is 26.1° C (Garcia, 1981).

## Alacranes Reef

Alacranes is an oval-shaped bank reef formation, covering an area of 293 $Km^2$ with a maximum length and width of 26.5 Km and 14.8 Km respectively (Figure 1J).

It is located on the north coast of Yucatan, 80 miles off the coast, between 22° 21' and 22 ° 35' N and between 89 ° 38' and 89 ° 49' W. The weather is hot and humid with summer and winter rains, average rainfall is 444 mm; the average temperature is 25.5° C (Garcia, 1981).

# RESULTS

The reproductive patterns that characterize the diversity of reproductive tactics are summarized below.

## *Mytella strigata* (Hanley, 1843) from Chautengo, Guerrero

- A large percentage of the population exhibited gametogenesis throughout the year (Figure 2) with peaks of 80% in February to March, 60% in June, and 80% in December.
- Two periods of maturity, one intense with 60% from April to June, and a minor one in September of 40%.

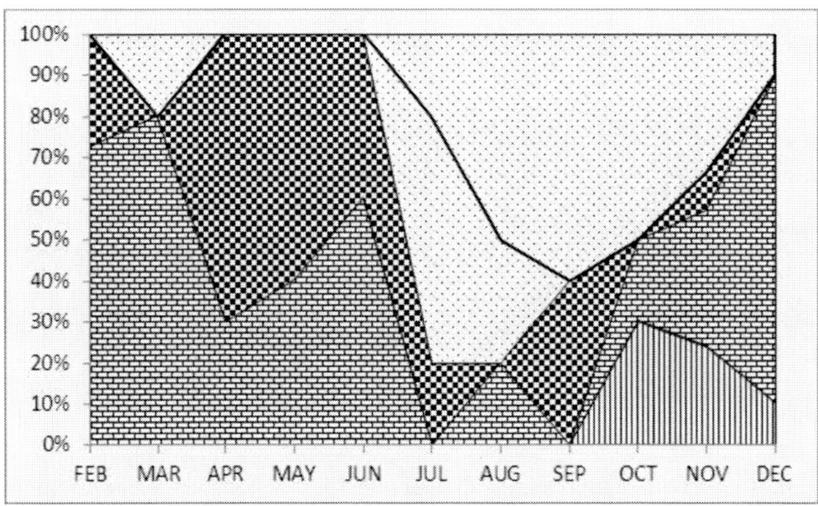


Figure 2. *Mytella strigata*. Reproductive cycle for both sexes from Chautengo lagoon, Guerrero.

- One well defined spawning period from July to September of 70%.
- Intense and clear post-spawn and rest periods from July to December with a total of 70%.

## *Mytella strigata* (Hanley, 1843) from Nuxco, Guerrero

- Constant gametogenesis through the year with a peak of 90% during June (Figure 3).
- One period of intense maturity from January to May.
- One well-defined spawning period from July to October with a maximum of 40%.
- Intense and clear post-spawn from September to December in 50% of the population and a rest period from February to June in 40%.

## *Chione undatella* (Sowerby, 1835) from La Paz, Baja California Sur

- Gametogenesis restricted to a low percentage of the population throughout the year (Figure 4).

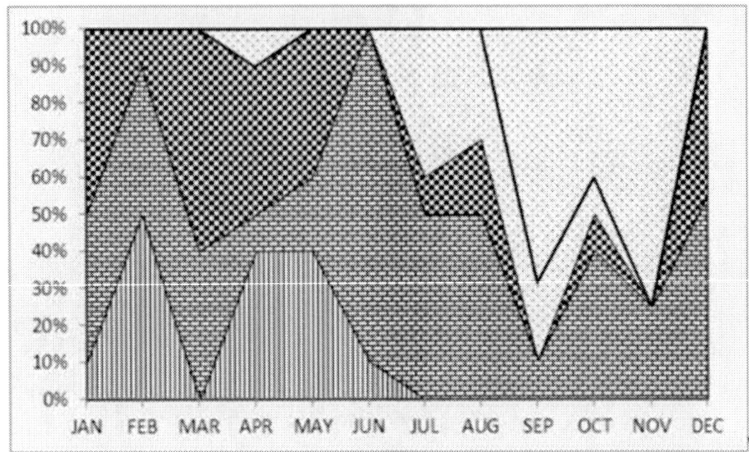


Figure 3. *Mytella strigata*. Reproductive cycle for both sexes from Nuxco lagoon, Guerrero.

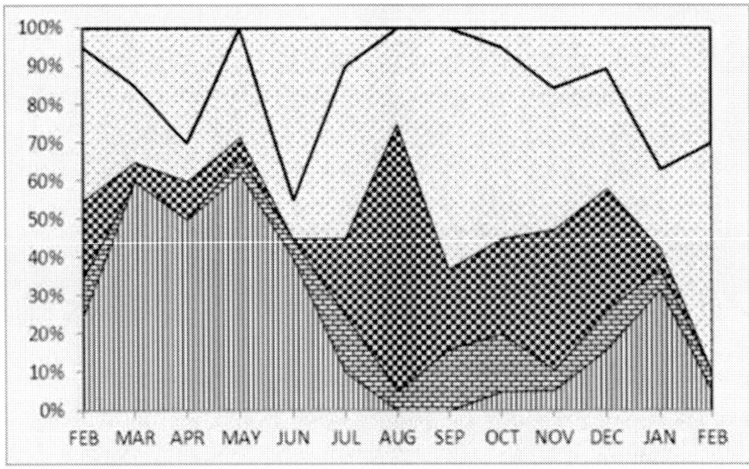


Figure 4. *Chione undatella*. Reproductive cycle for both sexes from La Paz, Baja California Sur.

- Constant maturity with a higher intensity from July to December and a maximum of 60% during August.
- Constant spawning with a maximum of 60% in September.
- Intense and clear post-spawn with three peaks during April, June, and January.

- Two rest periods, a broad one of 60% from March to June and a minor from December to February.

## *Anadara tuberculosa* (Sowerby, 1833) from La Paz, Baja California Sur

- Gametogenesis throughout most of the year with two peaks of high intensity in June (50%) and October (20%) (Figure 5).
- Maturity with 70% in February and an average of 20% through the year.
- Two high-intensity spawning periods that extend from February to May with a maximum of 60%, and from August to February with 85%.
- Post-spawning and rest periods occur throughout the year, with a maximum of a 30% post spawn during August.

## *Argopecten ventricosus* (Sowerby, 1842) from La Paz, Baja California Sur

- Intense gametogenesis in November and December with 70 to 100% of the population (Figure 6).

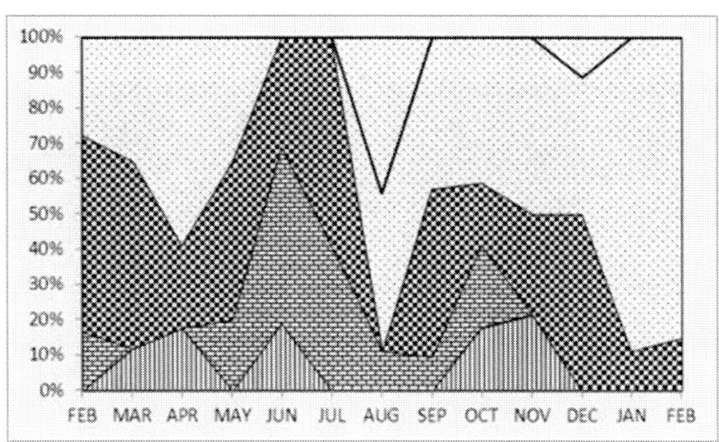


Figure 5. *Anadara tuberculosa*. Reproductive cycle for both sexes from La Paz, Baja California Sur.

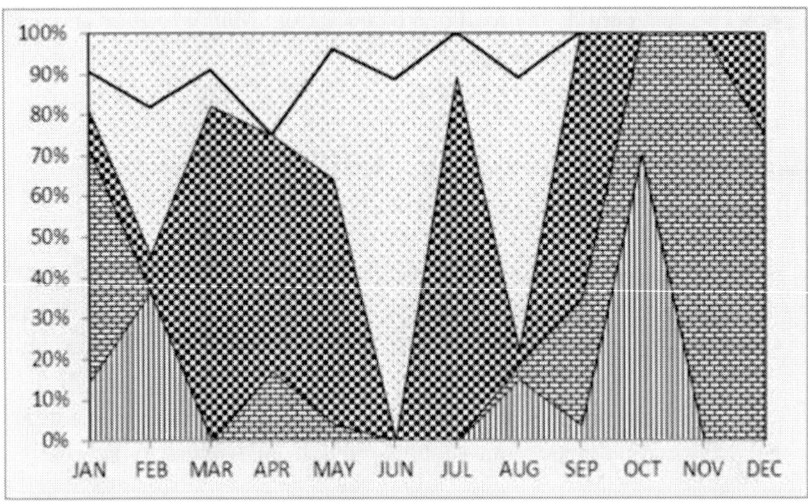


Figure 6. *Argopecten ventricosus*. Reproductive cycle for both sexes from La Paz, Baja California Sur.

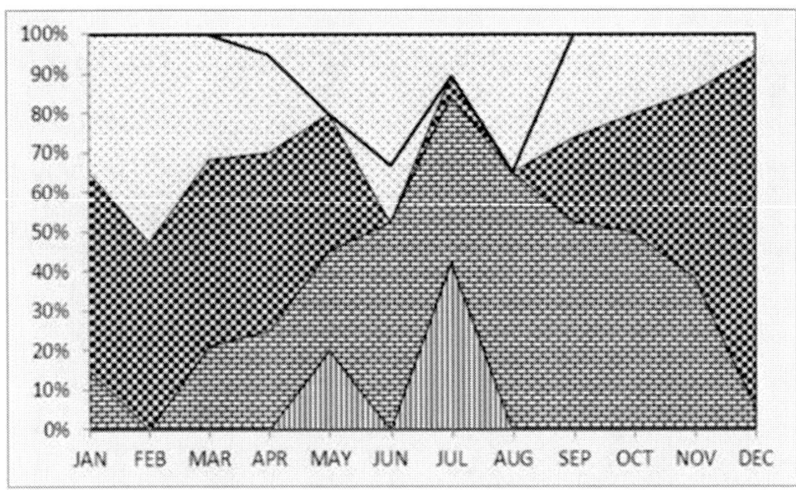


Figure 7. *Argopecten ventricosus*. Reproductive cycle for both sexes from Concepción Bay, Baja California Sur.

- Three periods of intense maturity from March to May, July, and September.
- Three high-intensity spawning periods with 30% in February, 90% in June, and 70% during August.

- Constant small percentage of post spawning and two well-defined rest periods during January to February and August to October with 30 and 70% of the population, respectively.

## *Argopecten ventricosus* (Sowerby, 1842) from Concepción Bay, Baja California Sur

- Gametogenesis represented by a high percentage of the population through the year with up to 70% in August (Figure 7).
- High percentage of mature organisms with up to 60% from September to May.
- Two spawning periods, a minor one of 15% in June and a major with 50% from September to April.
- Intense post-spawn and rest periods from April to July and again during May and June.

## *Argopecten ventricosus* (Sowerby, 1842) from Ojo de Liebre lagoon, Baja California Sur

- Intense gametogenesis limited to February to March with two small pulses during June to July and November to December (Figure 8).
- A constant high percentage maturity with up to 90% during November.
- Uninterrupted spawning with three low intensity peaks of 40% in March, August and December.
- Constant small percentage of post-spawn with no rest period.

## *Dosinia ponderosa* (Gray, 1838) from Zihuatanejo, Guerrero

- Gametogenesis throughout most of the year with three peaks in October, May, and August (Figure 9).
- Constant low percentage of maturity of no more than 30%.
- Uninterrupted spawning of 30 to 40% of the population with peaks of 50% during June and 60% in September.

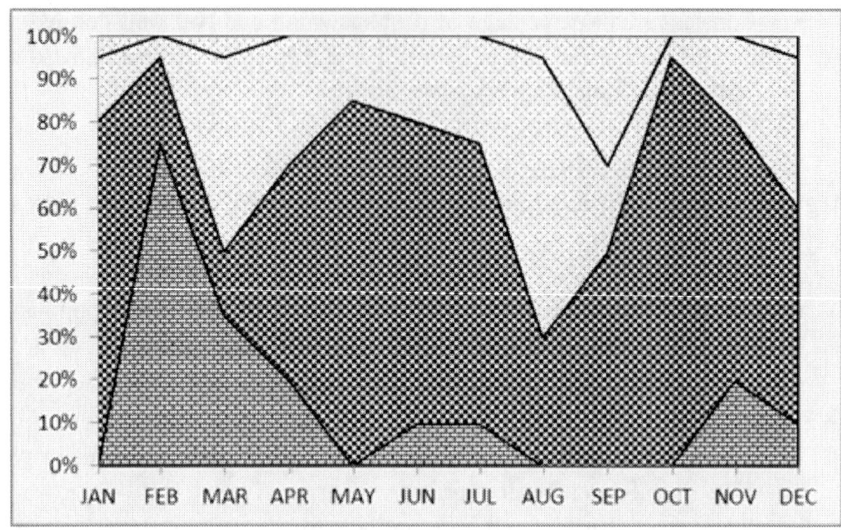


Figure 8. *Argopecten ventricosus*. Reproductive cycle for both sexes from Ojo de Liebre lagoon, Baja California Sur.

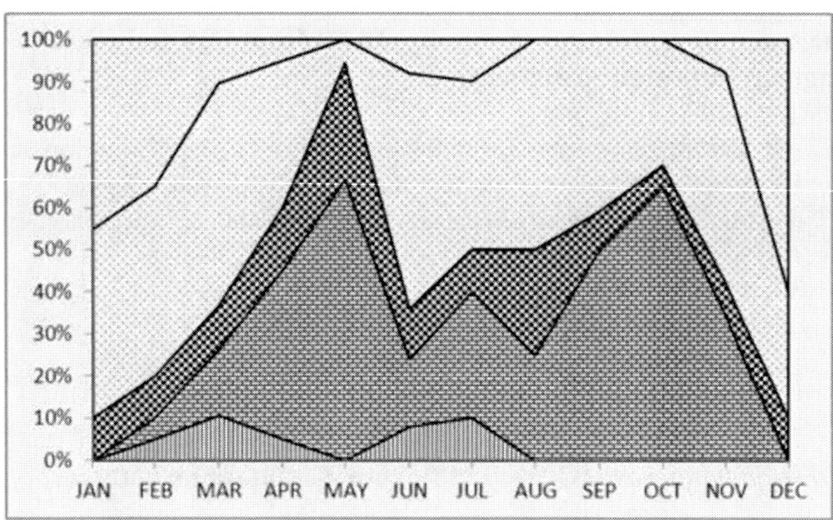


Figure 9. *Diosinia ponderosa*. Reproductive cycle for both sexes from Zihuatanejo, Guerrero.

- Clear post-spawn in December in 60% of the population with a minimum rest period from February to July.

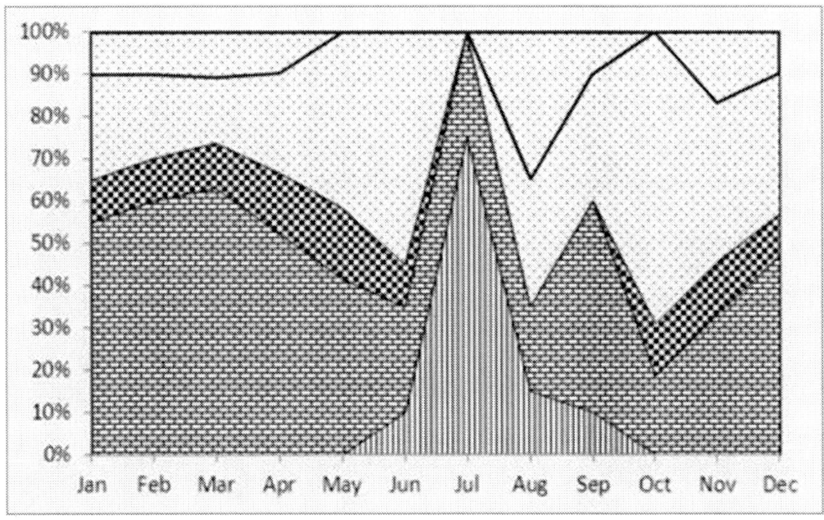


Figure 10. *Megapitaria squalida*. Reproductive cycle for both sexes from Zihuatanejo Bay, Guerrero.

## *Megapitaria squalida* (Sowerby, 1835) from Zihuatanejo, Guerrero

- Gametogenesis represented by a high percentage of the population from September to May with up to 60% during March (Figure 10).
- Constant low percentage maturity of no more than 10%.
- Constant spawning with peaks in October (70%) and June (50%).
- Post spawn is present through the year with a well-defined rest period from June to September and a peak of 70% during July.

## *Megapitaria aurantiaca* (Sowerby, 1831) from Zihuatanejo, Guerrero

- Gametogenesis was recorded throughout the year with peaks in November and March (Figure 11).
- Constant low percentage of mature organisms.
- Constant spawning of 5 to 20% of the population with a peak of 70% in October.

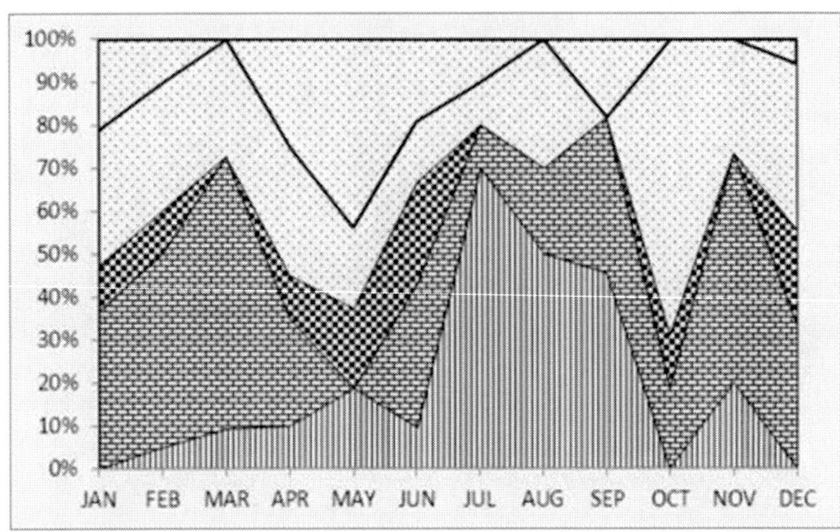


Figure 11. *Megapitaria aurantiaca*. Reproductive cycle for both sexes from Zihuatanejo, Guerrero.

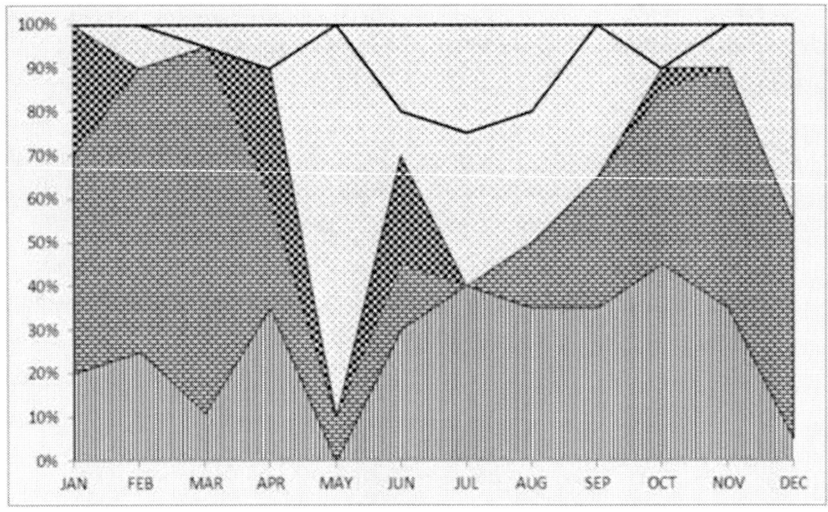


Figure 12. *Hexaplex erythrostomus*. Percent of organisms in the different reproductive stages from Conception bay, Baja California Sur.

- Post spawning in pulses with peaks during January and May, followed by organisms in rest phase with a peak of 70% during July.

**Chart 2. Reproductive Tactics of Bivalves**

| Gametogenesis restricted to a low percentage of the population through the year.<br>Gametogenesis represented by a high percentage of the population through the year.<br>Intense gametogenesis, high percentage, limited to a short period of time |
|---|
| Maturity |
| Lacking maturity or limited to a very low percentage of the population for a short period of time<br>Constant low percentage maturity<br>Constant high percentage maturity<br>Intense maturity previous spawning periods |
| Spawn |
| One well defined spawning period<br>Two or more low intensity spawning periods<br>Two or more high intensity spawning periods<br>Uninterrupted spawning |
| Post spawn |
| Short or absent post spawn<br>Constant small percentage of post spawned<br>Intense and clear post spawn |
| Rest |
| Few organisms in rest stage for a short time or absent<br>A clear well defined rest stage of low intensity<br>Rest stage present through the year of low intensity<br>A clear well defined rest stage of high intensity<br>Rest stage present through the year of high intensity |

## *Hexaplex erythrostomus* (Swainson, 1831). Conception Bay

This is a carnivorous species found on the same grounds as *S. gracilior*, but within a wider range of bottom types, from mud to gravel.

Figure 12 presents the percentage of both sexes in the different stages of the reproductive cycle from March 1979 to May 1980. A very significant percentage of organisms were at rest through the year, with up to 40% from July to November and 60% during May 1980. Gametogenesis was present in three periods with a higher percentage during March 1979 (80%), October (40%) and March and April 1980 (80%). Very short periods of maturity preceded the spawning periods, with a maximum of 30% during June 1979 and January and May 1980.

In Figure 12, four spawning periods can be identified, three for the year 1979 that extend from April to December, with three peaks, the highest of 90% during May and two secondary of 40% during July and December. Post-spawn is evident after the major spawning period in 20% of the population from June to August.

## *Strombus gracilior* Sowerby, 1825. Conception Bay

Organism for this study came from a population from Conception bay from sandy mud bottoms (Figure 13) at a depth of 3 to 15 meters.

Chart 3 presents the percentage of combined sexes in the different reproductive stages. A very high proportion of organisms at rest were present through the year, with peaks from February to April 1979 (> 60 %), November to December (> 80 %) and February 1980 (60%). Gametogenesis was present through the year with a clear period of dominance from May to October 1979, involving over 80% of the population during May and June (Figure 13). A corresponding proportion of mature organisms can be identified simultaneous to the gametogenic activity, but involving a lower number of organisms. Two spawning periods were detected, both involving a low proportion of the population. The first spawning period was at the beginning of both years from February to March 1979 and 1980 and the second was from August to November in only 20% of the population. A very high proportion (70%) of the population presented a post spawning stage, but only for a short time (December to March).

## *Strombus pugilis* Linné, 1758. Campeche Bay

The organisms for this study came from a population that extends from Seybaplaya to Campeche city (Figure14); it is a sub-littoral sandy area, 5 meters deep. Data for this species are sparse, covering April and June 1990 and from April 1996 to July 1997, but with a gap from October to January.

**Chart 3. Reproductive tactics presented by seven Gastropod populations of six species at four different localities**

| | Species | Locality |
|---|---|---|
| Gametogenesis | | |
| Intense gametogenesis | *Fasciolaria tulipa*<br>*Strombus gigas* | Campeche Bank, Campeche<br>Chinchorros reef, Quintana Roo |
| Low intensity gametogenesis | *Strombus gracilior*<br>*Hexaplex erythrostomus*<br>*Melongena corona* | Alacranes reef, Yucatan<br>Conception Bay, Baja Calif.<br>Conception Bay, Baja Calif.<br>Campeche Bank, Campeche |
| Spawning | | |
| Constant spawning | *Melongena corona*<br>*Fasciolaria tulipa*<br>Strombus pugilis | Campeche Bank, Campeche<br><br>Campeche Bank, Campeche |
| Two or more spawning pulses | *Strombus gracilior*<br>*Hexaplex erythrostomus* | Conception Bay, Baja Calif<br>Conception Bay, Baja Calif |
| One short spawning pulse | *Strombus gigas*<br>*Strombus gigas* | Chinchorros reef, Quintana Roo<br>Alacranes reef, Yucatan |
| Gonad recovery | | |
| Minimum or no post spawn and rest stages | *Strombus gigas*<br>*Hexaplex erythrostomus* | Chinchorros reef, Quintana Roo<br>Conception Bay, Baja Calif |
| Fast gonad recovery | *Fasciolaria tulipa* | Campeche Bank ,Campeche |
| Limited or no mature stage | *Hexaplex erythrostomus*<br>*Strombus gigas*<br>*Strombus pugilis*<br>*Melongena corona* | Conception Bay, Baja Calif<br>Alacranes reef, Yucatan<br>Campeche Bank, Campeche<br>Campeche Bank, Campeche |

A period of rest was detected during April and June of years 1990 and 1996 in up to 20% of the population. A very low frequency (5%) was also detected during August and September 1996 and a higher frequency of up to 30% during June and July 1997.

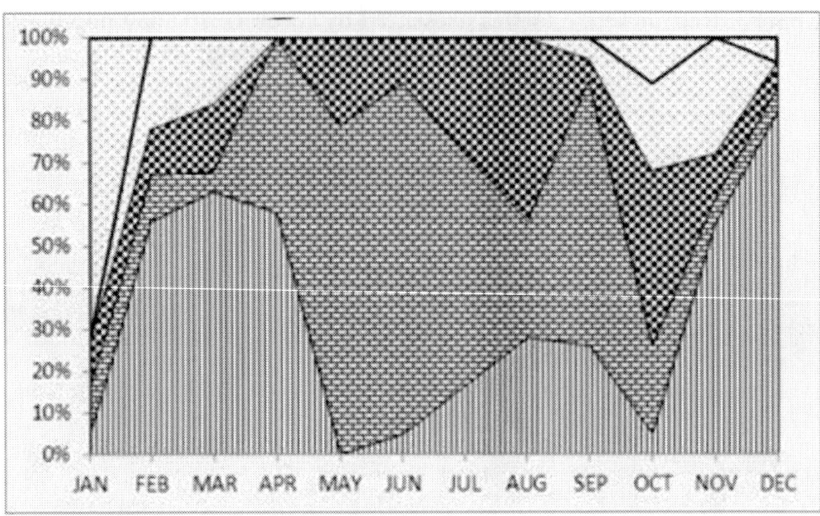


Figure 13. *Strombus gracilior*. Percent of organisms in the different reproductive stages from Conception bay, Baja California Sur.

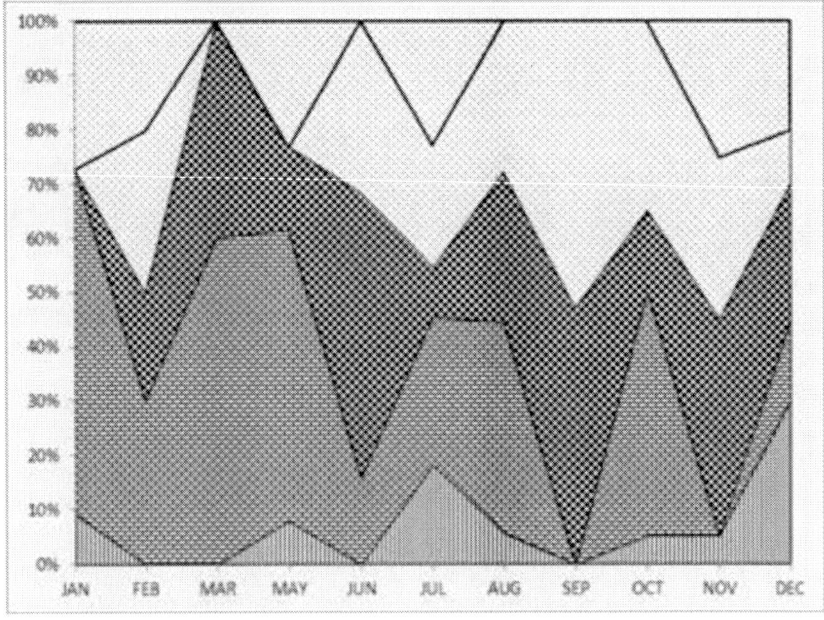


Figure 14. *Strombus pugilis* Percent of organisms in the different reproductive stages from Campeche bay, Campeche, Mexico.

Gametogenic activity and mature organisms were present in all samples (Figure 14), except for July 1996, during which only mature and spawning organisms were found.

Spawning organisms were found in June 1990, from June to September 1996, in February and from May to July 1997. Post spawn was detected during April and June 1990; April, June, and September 1996 and February and July 1997.

## *Melongena corona* (Gmelin, 1791), Campeche Bay

Species with a wide range of habitat preferences, from intertidal mangrove swamps and mud flats to sub-littoral sea grass beds. The individuals for this study come from sublittoral grass beds off the coast of Isla Arenas, on Northern coast of Campeche (Figure 15). Data for this species cover October 1999 to August 2000.

Two gametogenic pulses were detected, one during winter and spring with a maximum during December (63%) and another during summer with a peak during August (57%). Mature organisms were detected as early as December and were present until May with a maximum of 24% during February. Although spawn was registered from October to May, with peaks in January (56%) and April (68%), the presence of post spawners suggests a large percentage of organisms spawned during September to October (71%). The presence of a high frequency of rest organisms during July (63%), compared with the low percentage recorded after the post spawning pulse registered during October and November, suggests that the winter spawn is more demanding on the physiological condition of the species than the summer one (Figure 15).

## *Fasciolaria tulipa* (Linné, 1758), Campeche Bay

Organisms of this species came from intertidal mud and sea grass flats of Isla Arenas and Seybaplaya, Campeche (Figure 16). Sampling took place from September 1999 to August 2000.

Gametogenesis is present the year round, with dominance during spring and summer, peaks in February (69%) and June (65%), and the minimum during December (8%). Mature organisms are present the whole year, with the maximum in January (57%) and April (30%).

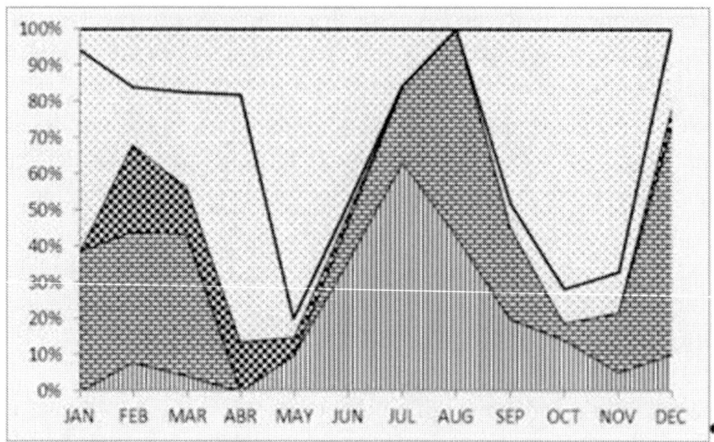


Figure 15. *Melongena corona* (Gmelin, 1791), Percent of organisms in the different reproductive stages from Campeche bay, Campeche, Mexico.

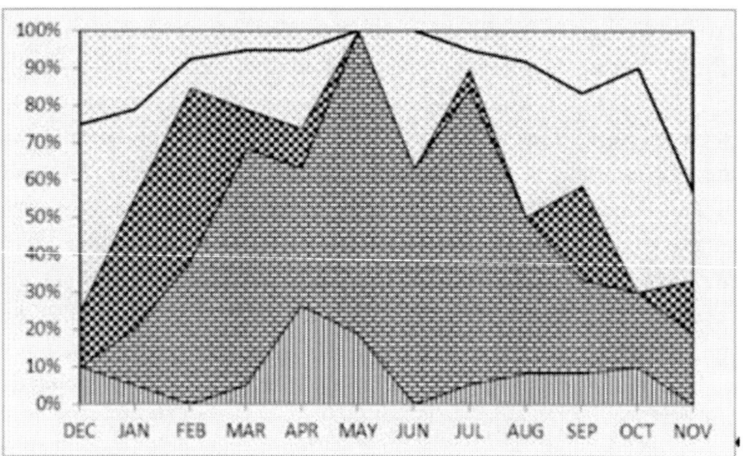


Figure 16. *Fasciolaria tulipa*. Percent of organisms in the different reproductive stages from Campeche bay, Campeche, Mexico.

Spawn also takes place year-round, with higher frequency during late summer (60%) and autumn (50%), but with significant peaks in spring (36%) and summer (25%). After every mayor spawning period, a significant frequency of post spawners and rest organisms were recorded with peaks in November (43%), May (32%) and July (51%) of the combined stages post spawn and rest (Figure 16).

## *Strombus gigas* Linné, 1758. Chinchorro reef, Quintana Roo

Organisms for this sample were collected from the central lagoon, from coral and sandy bottoms. Although the sampling period covers only from June to September, it was designed to cover the spawning period, which was suspected from the presence of egg masses in the field.

A rest period in 30% of the population was registered during September. Gametogenesis was present through the sampling period, with 40% of the population during June, declining gradually to a minimum in September (20%). A constant presence of mature organisms is evident in a percentage that fluctuated between 30 and 40% of the population, contrasting with the spawning period that apparently is limited to the summer with a peak during July with only 40% of the population. Only 10% post-spawn was detected during June (Figure 17).

## *Strombus gigas* Linné, 1758. Alacranes reef, Yucatan

*S. gigas* was sampled in the west of Perez Island, on sandy bottom. The sample for this population was also designed to cover the spawning period, based on the presence of egg masses in the field during previous studies.

A discontinuous rest stage was detected in up to 30% of the population, with only one peak correlated to the spawning period. Gametogenesis is constant through the sampling period, with a minimum of 30% during July and a maximum in May, prior to the accumulation of mature gametes that was present from June to September.

In the period that covers the spawning season, though, only a very low percentage of the population was caught during spawn; the very large percentage of post spawners clearly defines the spawning period from June to October (Figure 18).

# DISCUSSION

## Identification of Reproductive Patterns

From the analysis of the reproductive cycles, different patterns have been identified in every reproductive phase, depending on the percentage and duration.

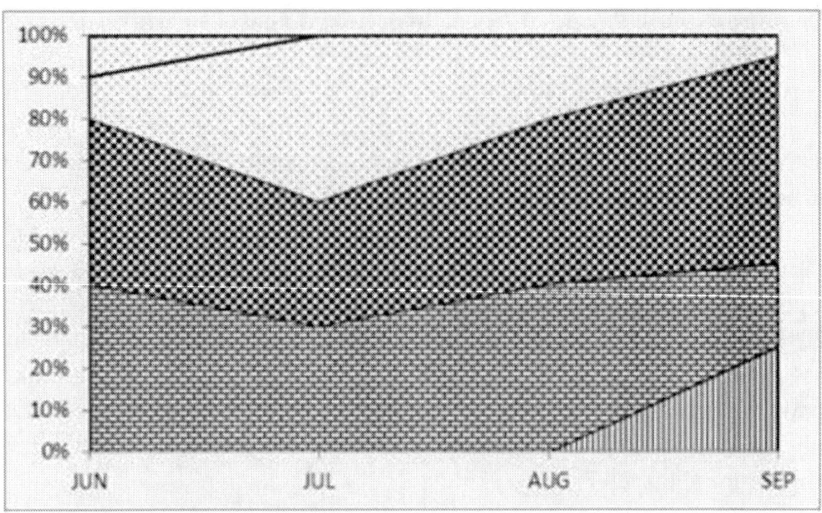


Figure 17. *Strombus gigas* Percent of organisms in the different reproductive stages from Chinchorro reef, Quintana Roo, Mexico.

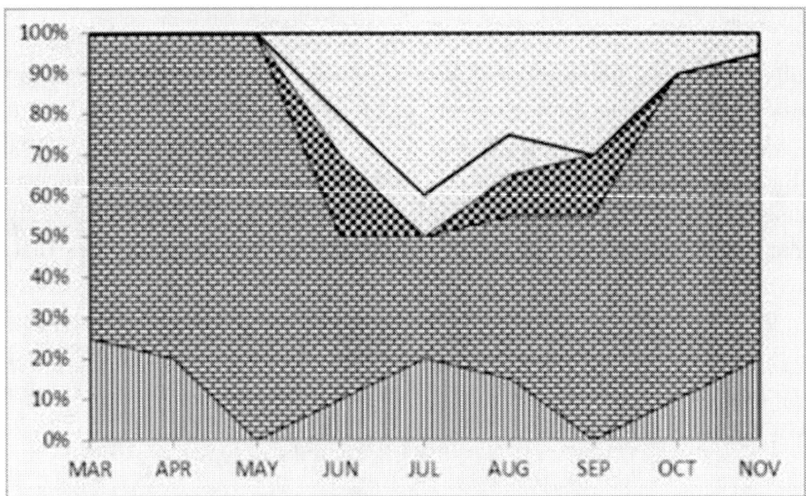


Figure 18. *Strombus gigas*. Percent of organisms in the different reproductive stages from Alacranes reef, Yucatan, Mexico.

For gametogenesis, a small percentage was interpreted as fast gametogenic activity in contrast to peaks of high percentage, which can be interpreted as slow gametogenic activity.

In maturity, a small percentage or absence through the year means there is no accumulation of ripe gametes, but as gametes become mature they are being shed. In contrast, high percentages of mature organisms either during short or long periods have been interpreted as the accumulation of ripe gametes.

The spawning phase is the product of both previous phases which can be a constant, low percentage, uninterrupted spawn, to one well defined period in a high percentage of the population. The higher the percentage during a shorter length of time, the more intense and synchronic the spawn.

The duration and intensity of post-spawn and rest phases are the result of the recovering capabilities of the populations, which is exemplified by some populations in that it can be different at different times of the year, and generally closely related to the intensity of the spawning phase.

Chart 12 summarizes the different reproductive patterns for every reproductive phase.

## Similarities and Differences among Populations of the Same Species

*M. strigata* seems to have a faster regeneration of gonadic tissue in the fall in the lagoons of Chautengo, which after spawning in summer gives way to slow gametogenic activity during winter and spring. While at Nuxco, the presence of a higher percentage of the populating in the rest stage during winter and spring suggests a slower gonad recovery, but faster gametogenic activity. Both produce a similar percentage of mature organisms, but with a difference in the intensity and duration of the spawning stage.

*A. ventricosus* had similarities in the maturity and spawning stages. The three populations revealed an accumulation of gametes throughout the year, previous to what seems to be three spawning pulses. These are clearly differentiated at La Paz, but less marked at Ojo de Liebre where they blend into a constant spawn throughout the year and yet are reduced at Concepción Bay, where the spawn at the beginning of the year is dominant with very marked differences in the recovering phases of the gonad and gametogenesis. At Ojo de Liebre, apparently no rest phase is required after any major spawn, with very fast gametogenesis during summer and fall. While at La Paz, a recovery period is required from fall to winter, with slow gametogenesis during that period but intense during spring and summer, whereas at Concepción Bay, gametogenesis is slow and constant through the year with a rest period during spring and summer.

## Similarities and Differences among Species from the Same Locality

The populations of *M. aurantiaca, M. squalida* and *D. ponderosa* share the same habitats at Zihuatanejo with a fine distribution determined by sediment characteristics (Baqueiro, 1979). The three populations exhibit a very similar behaviour in the maturity stage which is limited and constant through the year. Their spawning stage is constant with pulses of higher intensity in different months for each species. There were marked differences among species in the recovery of gonads and gametogenic activity. *D. ponderosa* seemed to have a limited rest stage during spring and summer while *M. squalida* had a clear rest period during summer, which was more intense for *M. aurantiaca* throughout the year. Gametogenesis is slow and occurs in pulses in *D. ponderosa* while in *M. squalida,* it is constant and slow during winter and spring but faster during the summer. It is similar in *M. aurantiaca* but is more intense and with less duration during fall and summer.

## *Chione undatella*, *Anadara tuberculosa*, and *Aechipecten circularis* from La Paz, Baja California Sur

Although these three species inhabit the same locality in the intertidal zone, they come from different habitats. *C. undatella* lives in sandy beaches, *A. tuberculosa* in mud among red mangrove roots, and *A. circularis* in sea grass beds of sandy mud. The three species differ in every stage of their reproductive cycle with *C. undatella* having the longest recovering period, but the fastest gametogenesis, and constant spawning with higher intensity during fall. *A. tuberculosa* exhibited a limited recovery period with slow gametogenesis during summer and faster in the fall with an extended reproduction period with peaks in spring, summer, and winter. While *A. circularis,* as mentioned before, had slow recovery and gonad activity during fall and winter and the opposite during spring and summer. The works of Webber (1977), Bayne (1978), Sastry (1979), Fretter (1984), and Mackie (1984) show that different environmental factors have diverse effects on gonad development, with temperature and food availability as the most important.

However, in environments such as tropical coastal lagoons, other factors may become critical, such as salinity, turbidity, and tidal exposure. Therefore, a gradient of "instability" as suggested by Lubet (1996) could be generated in coastal waters, which affect gonad development and other population parameters.

This "instability" gradient is reflected in the different expressions of the reproductive cycle, which given the reproductive flexibility, bivalves will try to optimize reproduction through one of several alternatives:

- Fast gametogenesis with accumulation of mature gametes.
- Slow constant gametogenesis with limited or no accumulation of mature gametes.
- Constant asynchronic spawning.
- Synchronic spawning in limited to extended periods of time.

This differs from the point of view of Lubet and Mann (1987), who identify two types of reproductive behaviour among bivalves: the "*Mytilus* type" with no temperature limits for gamete activity and spawning, and the "*Crassostrea* type," which requires a minimum temperature for spawning and an obligatory rest period.

From the results presented, it can be seen that the rest period is not a species-specific requirement but rather a response to environmental conditions as exemplified by *A. ventricosus*. In addition to temperature, some other factors may determine a rest stage as exemplified by salinity in *M. strigata*.

*Strombus gracilior* and *Hexaplex erythrostomus* are species with very different life strategies, the former an herbivore—detritivore, with a planctotrophic larval development—and the latter a predator and scavenger of direct development in an egg capsule. Both share the same habitat.

Although, a similar gonad behavior is evident—a prolonged rest period, followed by extended gametogenic activity and clear post spawn. Their spawning periods do not overlap.

*H. erythrostomus* has a very brief stage of gamete storage, with an almost continuous spawning, with peaks of maximum intensity and periods of minimum or null activity. Whilst *S. gracilior* presents a constant high percentage of the population with mature gametes, but only very brief spawning periods, that coincide with the periods of low or null activity of *H. erythrostomus*.

*Melongena corona*, *Fasciolaria tulipa,* and *Strombus pugilis*, although they live in the same locality, only occasionally, *M. corona* and *F. tulipa* can be found in the same habitat or *F. tulipa* with *S. pugilis*.

*M. corona* and *F. tulipa* are species with very similar life strategies, from a direct larval development to their feeding habits, with *F. tulipa* being restricted to the lower littoral and sub-littoral zone, while *M. corona* can be found from upper littoral to sub-littoral zones. The populations for this study come from the same habitats and present similar gonad behavior: constant gametogenic activity, accumulation of mature gametes, in particular during winter and early spring, and a year-round spawning, with distinct post spawn and rest periods. With variations on the intensity of gametogenesis, proportion of the population accumulating mature gametes, and a clear alternation of the spawning peaks. While *M. corona* had its peaks in January and April, *F. tulipa* presented spawning peaks in October, December, and two minors during March and June.

In contrast, *S. pugilis* with a different life strategy, from larval development to feeding habits, shares the same gonadic behavior, with constant gametogenic activity through the year, accumulation of mature gametes and post-spawn and rest periods. Differing on the time of the year the mature gametes are stored and limiting the spawning period to a single period during summer and autumn.

The previous examples show that the environment controls the behavior of the reproductive cycle, and that adaptive strategies are at work: (1) to minimize inter-specific competition in the case of *M. corona* and *F. tulipa* and (2) a predator-prey relationship for *S. gracilior* and *H. erythrostomus*.

For the two populations analyzed of *S. gigas*, although spawn occurs during the same months, July to August, significant variations are evident. The population from Chinchorro reef showed a constant presence of organisms with mature gametes, with up to 40% at the beginning of the spawning period, increasing up to 50% at the end of the spawning season, and a maximum percentage of spawners of 40%. With no evident post spawning, but a 10% at the beginning of the season. Whilst the population from Alacranes reef presented a preparation for spawn in just 20% of the population with the total participation of mature organisms at the peak of the spawning season, when up to 40% of the population have spawned and 10% were spawning. A similar increase in the percentage of mature organisms was also detected at Alacranes at the end of the spawning season, but it had totally disappeared in October. This population presented a clear post-spawn and fast gonad recovery, represented by the high percentage of organisms in gametogenesis during

September. Apparently in both populations, a period of re-absorption started in autumn, represented by the increasing percentage of the rest stage after September.

Being that Chinchorro bank is a more stable environment, more organisms present a readiness for spawning, but fewer spawn. Reproductive seasons reported for *S. gigas* in the Caribbean region does not include histological results. The reproductive season of *S. gigas* has been obtained by observations of copulating or egg-laying periods. Stoner *et al.* (1992) make recompilation of the reproductive seasons for queen conch based in reproductive behavior, taking data from different authors. At Bermuda (the most northern locality), the reproductive season for *S. gigas* began in May and ended in September. In the Florida Keys, the reproductive season was between middle January to middle September. At Venezuela (the most southern locality), the reproduction was from middle March to mid November. Corral and Ogawa (Stoner et al., 1992) showed that *S. gigas* at Chinchorro bank, Quintana Roo, Mexico egg-laid year-round. Pérez Pérez and Aldana Aranda (2000) reported a reproductive period of *S. gigas* from May to September in Alacranes reef. Stoner et al. (1992) said there was no apparent trend related to latitude in beginning, end, or length of reproductive season for the queen conch from Bermuda to Venezuela. Geographic comparisons of seasonal reproduction must be interpreted cautiously due to different methods, frequency, and number of observations and different habitat types. In this study the gametogenic cycle for *S. gigas* presented differences, even though the gonads samples were taken during the same period for Alacranes reef and Chinchorro bank. Therefore, it can be ascertained that the reproductive activity was controlled by the environmental conditions.

*S. pugilis* and *S. gracilior* are corresponding species on different oceans, with the same type of development, habitat preference and feeding habits. Even though their reproductive cycle behaves in a similar way: constant gametogenesis, store of mature gametes and a post spawn and rest stages. Spawn for *S. gracilior* is restricted to two short periods, while *S. pugilis* covers a wider season and a higher population number, with an ample rest period for the population of *S. gracilior*.

Several authors have shown that duration and intensity of the gonad stages are a function of temperature and food availability (Bayne, 1978; Sastry, 1979; Webber, 1977; Fretter, 1984; Mackie, 1984), as well as other environmental factors (Bricelj and Malouf, 1980; Syed-Shafe, 1980; Jones, 1981; Kennedy and Krantz, 1982; Shephton, 1987; Brevber, 1980; Jaramillo *et al.* 1993).

In the reproductive cycle, the factors that induce gametogenesis and define the length and intensity of recovery periods are temperature and food availability (Lubet and Choquet, 1971; Bayne, 1978). Spawn may be induced by several factors such as temperature (Galtsoff, 1954; Holand and Chew, 1973; Hines, 1979), salinity fluctuations (Cain, 1974; Stephner, 1981), currents (Davis and Chanley, 1955; Ino, 1970), the presence of some micro algae (Brese and Robinson, 1981), among others.

## CONCLUSION

The gonadic cycle of any one species may vary in the presence, duration, and intensity of the rest and post-spawn, as well as the duration and intensity of the gametogenic, maturity, and spawning periods.

Two gametogenic strategies were identified as a response to the environment: (1) populations with a short gametogenic stage or a small percentage of gametogenic organisms, which represents fast gametogenesis (*F. tulipa and S. gigas* from Chinchorros); (2) Populations with a continuous gametogenesis in a very large number of organisms (*S. gigas* from Alacranes, *H. erythrostomus, S. gracilior,* and *M. corona*).

In regards to spawning intensity and duration, three variants were identified: (1) populations with one very extended spawning period, with or without a dominant peak (*Melongena corona, Fasciolaria tulipa,* and *Strombus pugilis*); (2) Populations with two or more clear peaks or spawning pulses (*Strombus gracilior* and *Hexaplex erythrostomus*) and (3) Populations with one short defined pulse (*Strombus gigas* from both localities).

The capacity of gonad regeneration is another factor that expresses itself in the duration and intensity of the post spawning and rest periods, and on the gametogenic and maturity periods. For this, three variants were identified: (1) Populations without or with very short and low-intensity post spawning and rest periods (*S. gigas* from Chinchorro reef and *H. erythrostomus* from Conception bay); (2) Populations with fast gonad recovery, in which only a small percentage of organisms can be detected in gametogenesis and with a clear and constant presence of mature gametes (*Fasciolaria tulipa* from Campeche Bank), and (3) Populations with limited or no mature stage that can support a constant or intense spawn. The characteristics presented are summarized in Chart 2 for bivalves and Chart 3 for gastropods.

When a competitive pressure, *M. corona* and *F. tulipa* from Campeche bank, or predatory pressure, *S. gracilior* and *H. erythrostomus* from

Conception Bay, were detected, a tendency for shorter, more intense, and alternating cycles was evident.

## REFERENCES

Baqueiro, C. E. 1979. Sobre la distribución de *Megapitaria aurantiaca* (Sowerby 1831), *M. squlaida* (Sowerby 1835) and *Dosinia ponderosa* (Gray 1838) en relación a la granulometría del sedimento (Bivalvia: veneridae): Nota científica. An. Cent Cienc. del Mar. and Limnol. Univ. Nat'l. Autón., México. 6: 25–32.

Baqueiro, C. E. 1981. Reproductive cycle of six species of bivalve mollusks with aquaculture potential, from the Panamic province. *World Mar. Soc.,* (Resumen): 367.

Baqueiro, C. E. and M. Castagna. 1988. Fishery and culture of selected bivalves in México: past, present and future. *J. Shellf. Res.,* 7: 433–443.

Baqueiro, C. E. and J. Stuardo. 1977. Observaciones sobre la biología, ecología y explotación de *Megapitaria aurantiaca* (Sow., 1931), *M. squalida* (Sow.) *y Dosinia ponderosa* (Gray, 1938) de Zihuatanejo e isla Ixtapa, Gro. , México. An. Cent. Cienc. del Mar y Limnol. *Univ. Nat. Autón.*, México. 4: 161–208.

Bayne, B., 1978. Reproduction of Bivalve mollusks under environmental stress. En: Physiological ecology of estuarine organisms. J. F. Vernberg (ed.) the Belle W. *Baruch library in marine science*, (3): 259-278.

Berg, J. C., and N. L., Adams., 1984. Microwave fixation of marine invertebrates. *J. Exp. Mar. Biol. Ecol.,* 74: 195-199.

Breese, W. P. and A. Robinson, 1981. Razor clams, *Siliqua patula* (Dixon): gonadal development, induced spawning and larval rearing. *Aquacultre,* 22: 27-33.

Brevber, P. 1980. Annual gonadal cycle in the carpet shell clam *Venerupis decussata* in Venice lagoon, Italy. *Proc. Nat. Shellf. Ass*., 70 (1): 39-85.

Bricelj, M. V. and R. E. Malouf, 1980. aspects of reproduction of hard clams (*Mercenaria mercenaria*) in Great South bay, New York. *Proc. Nat. Shellf. Ass.,* 70: 216-229.

Cain, T. M. 1974. Reproduction and recruitment of the brackish water clam *Rangia cuneata*. *Fish. Bull.,* 73 (2): 412-430.

Castagna, M. and J. N. Kaeuter, 1981. Manual for growing the hard clam *Mercenaria. Sp. Vir. Inst. Mar. Sci. Rep. Apl. Mar. Sci. and Ocean Enger.,* 249 : 110 pp.

Davis, H. C. and P. E. Chanlely, 1955. Spawning and egg production of oysters and clams. *Proc. Nat. Shellf. Ass.,* 46: 40-51.

Fretter, V. 1984. Prosobranchs. En: The Molluska, Wilburn, M K. (Ed.), Vol 7. Reproduction, Tompa, A. S., N. H. Verdonk and J. A. M. den Biggelaar (Eds.). Academic Press, Orlando Florida, U. S. A. Cap. 1: 1-45.

Galtsoff, P.S. 1964. The American oyster *Crassostrea virginica* Gmelin. *Fish. Bull.*, 64:1-480.

García, E., 1981. Modificaciones al sistema de clasificación climática de Köppen. Ed. Larios, México, segunda ed. D.F.: 252 pp.

Graffé, F. S., 1990. Tablas de Predicción de Mareas. Puertos del Golfo de México y mar Caribe. Datos Geofísicos serie A Oceanografía. *Inst. Geof. U.N.A.M.*: 113 - 144.

Hines, A. H., 1979. Effects of a thermal discharge on reproductive cycles in *Mytilus edulis* and *Mytilus californianus* (Molluska, bivalvia). *Fishr. Bull.,* 77 (2): 125-132.

Holland, D. A. and K. K. Chew. 1973. Reproductive cycle of the manila clam (*Venerupis japonica*) from hood canal, Washington. *Proc. Nat. Shellf. Ass.*, 64: 65-72.

Ino, T., 1970. Controlled breeding of mollusks. Indo Pacific Fishr. Coun. F. A. O. IPFC/070/SYM 35: 24 pp.

Jaramillo, R., J. Winter, J. Valencia and A. Rivera. 1993. Gametogenic cycle of the Chiloe scallop (*Chlamys amandi*). *J. Shellf. Res*., 12: 59–70.

Jones, S. D., 1981. Reproductive cycle of the Atlantic surf clam *Spisula solidissima*, and the ocean quahog *Artica islandica* off New Jersey. *J. Shellf. Res.,* 1 : 23-32.

Kanti, A., P. B. Heffernan and R. L. Walker. 1993. Gametogenic cycle of the southern surfclam, *Spisula solidissima* similis (Say, 1822) from St. Catherines sound, Georgia. *J. Shellf. Res.,* 12: 255–261.

Kennedy, V. S. and L. B. Krantz, 1982. Comparative gametogenic and spawning patterns of the oyster *Crassostrea virginica* (Gmelin) in central Chesapeake bay. *J. Shellf. Res*., 2 (2): 133-140.

Knaub, R. S. and A. G. Eversole, 1988. Reproduction of different stocks of *Mercenaria mercenaria. J. Shellf. Res*., 7 (3): 371-376.

Lankford, R. R., 1974. Coastal lagoons of Mexico, their origin and classification. En: Estuarine processes. Circulation, sediments and transfer of material, in the Estuary, Cronnin, L. E. (Ed), Academic Press Inc. New York, 2: 182 - 215.

Loosanoff, V. L. and H. C. Davis, 1963. Rearing of bivalve mollusks. In: Advances in Marine Biology. Russell, F. S. (Ed) Academic Press, New York 1: 2 - 136.

Lubet, P., 1996. Reproducción de los moluscos. Pages 171–216 in G. Barnabé, ed. Bases biológicas y ecológicas de la acuicultura. Trad. E. Cunchillos. Ed. Acribia S. A. España.

Lubet, P. and C. Choquet, 1971. Cycles et rythmes sexuales chez les mollusques bivalves et gastropods. *Haliotis*, 1 (2): 129-149.

Lucas, A. 1965. Recherche sur la sexualité des mollusques bivalves. Tesis doctoral Fac. Sci. Univ. Rennes, Francia: 136 pp.

Luna, G. L., 1968. Manual of Histologic staining methods of the Armed Forces Institute of Pathology. Tercera Ed. McGraw-Hill Book Co. New York: 258 pp.

Mackie, G. L., 1984. Bivalves. En: The Molluska. Wilbur, K. M. (Ed). Vol. 7. Reproduction Tompa, S. A., N. H. Verdonk and J. A. M van den Biggelaar, Ac. Press. Londres, U. K. Cap. 5: 305-418.

Pérez Pérez M. and Aldana Aranda D. 2000. Reproductive activity of Strombus gigas (Mesogasteropoda: Strombidae) in differente habitats of Alacranes reef, Yucatán. Revista de biologia tropical (Impact Factor: 0.46). 07/2003; 51 Suppl 4:119-26.

Phleger, F. B. 1965. Sedimentology of Guerrero Negro Lagoon, Baja California, México. Colston Res. Soc. Lond.: 200–235.

Sastry, N. A., 1979. Pelecypoda (Excluding Ostreidae). En: Reproduction of Marine Invertebrates, Giese, A. C. and J. S. Pearse, (Ed), Academic Press, Vol. 5: 113 - 292.

Shephton, T. W., 1987. The reproductive strategy of the Atlantic sur clam, *Spisula solidissima,* in Prince Edwards island, Canada. *J. Shellf. Res.*, 6 (2): 97-102.

Stephner, D., 1981. Induction of Spawning four species of Bivalves of the Indian coastal waters. *Aquaculture*, 25: 153-159.

Stoner, A. W., J. S. Veronique, and I. F. Boidron-Metairon. 1992. Seasonality in reproductive activity and larval abundance of queen conch Strombus gigas. *Fish. Bull.*, 90: 161–170.

Syed-Shafe, M. 1980. Quantitative studies on the reproduction of black scallop, Chlamys varia (L.)

Syed-Shafe, M., 1980. Quantitative studies on the reproduction of black scallop, *Chlamys varia* (L.) form Lanveoc area (Bay of Brest). *J. Exp. Mar. Biol. Ecol.*, 42: 171-186.

Villarroel, M. 1975. Relación entre macro-invertebrados bentónicos (especialmente moluscos) y sedimentos en tres lagunas del estado de Guerrero, México. Cent. Cienc. del Mar y Limnol. U.N.A.M. 64 p.

Webber, H. H., 1977. Gastropoda: Prosobranchia. En: Reproduction of Marine Invertebrates. Vol. IV. Mollusks: Gastropods and Cephalopods. Giese, C. G. and J. S. Pearse (Eds.) Ac. Press. Londres, U. K.: 1-98.

# EDITOR'S CONTACT INFORMATION

**Dr. Erick R. Baqueiro Cárdenas**
Federico Chopin 68,
Xalapa, Veracruz 91190,
Mexico
Tel: +52 228 812 8755
E-mail: ebaqueiro@gmail.com

# INDEX

## A



## B



## C

## D



## E

## F



## G



## H



## I

## J



## K



## L



## M

## N



## O



## P

## R



## S

## T



## U



## V



## W

## X



## Y



## Z